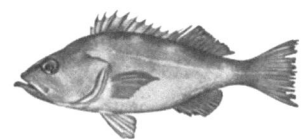

THE RISE & FALL
of the
PACIFIC OCEAN PERCH

*The History of the
R/V John N. Cobb and
Exploratory Fishing
(1950-1970)*

CHARLES ROBERT HITZ

The Rise & Fall of the Pacific Ocean Perch

ISBN (hardcover): 978-0-9720255-5-3
ISBN (paperback): 978-0-9720255-3-9
ISBN (ebook): 978-0-9720255-4-6
Library of Congress Control Number: 9780972025539

Published by Sitka 2 Publishing
www.sitka2.com

Copyright © 2025 Charles Robert Hitz. All rights reserved. No part of this work may be reproduced or transmitted in any form or by any means, electronic or mechanical, including photocopying and recording, or by any information storage or retrieval system, except as may be expressly permitted by the 1976 Copyright Act or in writing from the publisher. Requests for permission can be emailed to goaskchaz@gmail.com.

Cover images from various contributors: *Sebastes alutus*, NOAA FishWatch, Public domain, via Wikimedia Commons; Admiralty Routeing Chart, United Kingdom Hydrographic Office, Public domain, via Wikimedia Commons; Photos of R/V *John N. Cobb* by Charles Robert Hitz.

Editorial services by Misti Moyer, MistiMoyer.com
Design by Monica Thomas for TLC Book Design, TLCBookDesign.com

*I dedicate this book to
Scott Collins and Savonnah Mitchell.
When they graduated from high school
they were focused on science.*

TABLE of CONTENTS

Introduction 7

Prologue 9

CHAPTER 1   R/V *John N. Cobb* 1950–1960 13

CHAPTER 2   Lee Alverson's Interest 1949–1959 37

CHAPTER 3   College Years 1951–1960 51

CHAPTER 4   Cruise 46, Spit 1960 61

CHAPTER 5   My First Trip, Cape Scott, BC 1960 73

CHAPTER 6   My Second Trip, Heceta Bank, OR 1961 ... 91

CHAPTER 7   Between Trips 105

CHAPTER 8   Gulf of Alaska Cruises 1961–1962 115

CHAPTER 9   Columbia River (AEC) Cruises 1962–1963 .. 127

CHAPTER 10   Pacific Hake 1964–1966 149

CHAPTER 11   Russian Fleet 1966–1970 159

CHAPTER 12   *Seafreeze Pacific* & *G. B. Reed* 171

CHAPTER 13   NOAA Formed 1970 195

Acknowledgments 205

Appendices 207

References 217

Image Credits & Attributions 221

Afterword 225

About the Author 227

Introduction

In August of 1960, a newly hired biologist, on his first trip on the exploratory research vessel *John N. Cobb,* hiked to the top of Triangle Island, British Columbia, with an older crew member. At that moment, he knew he had made the right decision in taking the job with the US Fish and Wildlife Services, Exploratory Fishing and Gear Research Base. What he had observed in the catches made by the vessel so far was fascinating. He was in the right spot to continue gathering knowledge on this large group called rockfish (*Sebastes*).

The vessel was taking a break from exploring for new trawl hauls in areas reportedly untrawlable. Because of the exceptionally good weather, the vessel anchored close to the island. She was protected from the northwest swell, and the crew had the opportunity to walk the beach.

Arnie, a fisherman and big man, and a young biologist decided to climb to the top of the island and inspect the abandoned lighthouse. Arnie broke the trail up the side of the mountain, with the biologist following in his wake like a bear cub following its parent. Once they reached the lighthouse, they looked out at the vast expanse of the Pacific Ocean, spreading out before them to the horizon. Looking down the path they came up, they could see the white speck of the *Cobb*, which made them feel so small!

The biologist looked at the vast expanse of the ocean and wondered what lived under the surface. He was especially curious about one species of rockfish, *Sebastes alutus,* the Pacific Ocean perch (POP). How did it, one of a group of over fifty species, become so popular?

Prologue

I was the biologist standing on the top of Triangle Island and looking out over the Pacific Ocean. I realized I had the opportunity to understand where POP lived in the deep water of the continental slope, below 100 fathoms, where the slope begins. I didn't know much about what species of rockfish lived there or how deep the POP went. One of the jobs I was hired for was analyzing the catch taken from a commercial trawl that fished along the ocean bottom at different depths and to see what species lived at those depths. The hauls were made by the Exploratory Fishing Group.

A commercial trawl or otter trawl is a net that fishermen used to capture fish by towing it along the bottom. The net is made up of three parts: the cod end, intermediate, and two wings. A head rope extends across the top of the net from one wing to the other, with floats attached to it that hold the top of the net open while it is towed. Below there is a weighted foot rope that extends from each wing and holds the bottom of the net open next to the bottom. The net is spread by a pair of trawl doors that are each attached to the net wings by bridles. The doors act as kites when towed through the water, each shearing to an opposite side of the boat and spreading the net so it is open.

The depth of a series of ten hauls the *Cobb* made while I was aboard, before our venture to the top of Triangle Island, gave me the opportunity to see how the different species changed as we went deeper. They made the series by chance from shallower to deeper starting at 50 fathoms (300 feet) and down in increments to 120 fathoms (720 feet). Using the trawl net was no different from what we had done in high school. We towed a small plankton net behind a skiff in the San Juan Islands during the summer to see what was caught in it. The catch was plankton, fish eggs, and larvae, the difference being that the trawl was much larger, so we could control the depth and catch fish. Trawl fishermen have learned by experience at what depths marketable species can be taken. They knew that POP was found near the continental break (100 fathoms) but wanted to know how much deeper they extended.

I saw my first freshly caught POP when we made the first set with an otter trawl. The haul was made at a depth of 85 fathoms (510 feet), and the total catch, estimated at 2,000 lb., was sorted into four groups: flatfish, round fish, rockfish, and scrap fish. I was excited to find a variety of rockfish in the group, which was again broken down into two additional groups: black and red rockfish. There were five species of rockfish all mixed together—one species of black, the silvergray rockfish, and four red, canary rockfish, flag rockfish, Bocaccio, and POP (Fig. 1). Once they were separated, recognized as a species, and finally identified, there were approximately 500 lb. of POP. Appendix 1 lists all the rockfish species that appear in this book by their scientific names as well as their common names.

Fig. 1 Pacific Ocean Perch (POP), *Sebastes alutus*. Image #3731.

I was extremely excited to actually see and examine a freshly caught POP. I had seen pickled specimens at the university and specimens in the fish house as they were offloaded from the boats. They were carried fresh in the fish holds with ice to help preserve them. By the time they arrived, all their colors were faded. The museum specimens were all bleached out due to preservatives in which they were submerged. In the fresh samples that I inspected, the color was still bright and slowly fading. To be able to see freshly caught fish and be involved in the expanding POP fisheries was the job I wanted. I was fortunate to get hired in 1960 at this exciting time in the history of the northwestern fishing industry, and in such a dynamic group as the Exploratory Fishing and Gear Research Base, headed by Dayton L.

Alverson. They had their own research vessel, the *John N. Cobb* (Fig. 2), a 93-foot vessel built for exploratory fishing and crewed by fishermen who knew the industry and how to fish. Thus began my exciting and rewarding career as a marine biologist.

Fig. 2 United States Exploratory Research Vessel (R/V) *John N. Cobb* (1950). Image #3705.

This book is a story of some of my experiences in exploring the northeastern Pacific Ocean for new fisheries resources. It is also a story of my interest in one species of rockfish, *Sebastes alutus*, whose market name is Pacific ocean perch (POP). Once I got involved with the exploratory vessel *John N. Cobb,* I wanted to know about how she became an exploratory vessel. When I was hired in this group, I felt there was a need to save the exploratory records after it was abolished. I always felt, after the first three exploratory cruises that I was on and involved in writing the final report, that we had left something out. We recorded the presence of each species of rockfish, but how many of each did we catch compared to POP? Why did the new factory stern trawler have such an effect on the POP resource? I hope my story will stimulate students to consider science as an exciting career. I also hope they will learn something about the ocean, fish biology, and the history of exploratory fishing.

CHAPTER ONE

R/V *John N. Cobb* 1950–1960

Fig. 3 R/V *John N. Cobb* (1960) — Rigged as a Trawler. Image #3763.

The Exploratory Fishing and Gear Research Base was located in the US Fish and Wildlife Montlake Laboratory brick building located just south of the Montlake Bridge in the University District of Seattle in 1960. This was the same building that became the Northwest Fisheries Science Center when the National Oceanic and Atmospheric Administration (NOAA) was formed in 1970. In 1960, there were three groups in the facility: Biological Laboratory, Technological Laboratory, and Exploratory Fishing and Gear Research Base. On July 5, 1960, I walked into the building and reported to work at my new job, which I believed was a biologist for the Exploratory Fishing group.

While waiting in the hall on the second floor of the building, I met Al Pruter, who was also waiting to get fingerprinted and sworn in as a new hire like me. Al was another WWII veteran like Lee Alverson, who became our boss, since he had recently become the new director of the Exploratory Fishing and Gear Research Base. Lee and Al were classmates at the University of Washington (UW) College of Fisheries, which was renamed in 2000 to UW Aquatic and Fishery Sciences. Lee hired Al as his deputy director. Al was an excellent administrator, and they made a wonderful team. Lee was good with groups, while Al was better with individuals and was outstanding in writing and planning. Quiet and steady, Al would be an important mentor in my career.

I was anxious to go aboard and inspect the *John N. Cobb* (Fig. 3). I had heard so much about her during my time at the College of Fisheries. The vessel had been designed and built in 1950 solely for exploratory fishing.

My First Inpection

Al took me to the *Cobb* to inspect her about a week later.[1] We went down to the dock on Lake Union at the foot of Stone Way in the University District on a beautiful sunny day for which Seattle is famous in the summer. As we approached, we could see the *Cobb's* cruiser stern reflecting the sunlight off her paint, with her full name across the stern and underneath — "F.W.S. 1601," the vessel numbering system of the US Fish and Wildlife Service. She looked as if she was newly commissioned, like a well-kept, high-liner fishing vessel that the owners took pride in, yet she'd been in service for almost ten years.

From the outside, the *Cobb* looked like a typical Pacific coast sardine seiner, a large combination vessel (Drawing 1). Her total length was ninety-three feet with a twenty-five-foot beam, built of wood by the Western Boat Building Company, a Tacoma shipyard with years of experience in crafting wooden vessels to withstand the Pacific Ocean. She was husky for a fishing boat, and other differences became apparent as we approached. The top rail surrounding the main deck was abnormally wide due to the added strength of wide ribs. Ironbark, a hard, durable wood, could be seen just above the waterline. It extended out by an inch and covered the hull from above to below the waterline, protecting the hull from being cut in two when going through sheet ice. She was expected to work in the ever-changing Pacific weather and had been doing so for the past ten years, exploring waters from California to the Bering Sea after she was built in 1950.

She was rigged as a trawler. A wide fish deck and open space made it ideal for trawling. Flat trawl doors were stored on the outside of the rails, attached to trawl stanchions bolted to the afterdeck. The tackles, single 1 and 2, and the double rigged to the boom were used to bring in the trawl net and its catch aboard after the trawl was towed along the bottom. The catch would be dumped into the checker where I would help sort out the catch by species. She was a typical Pacific combination vessel, which meant she could easily be converted to other fisheries. We boarded her on the fish deck and headed for the Dutch door on the back of the house, which led into the galley, the heart of the ship.

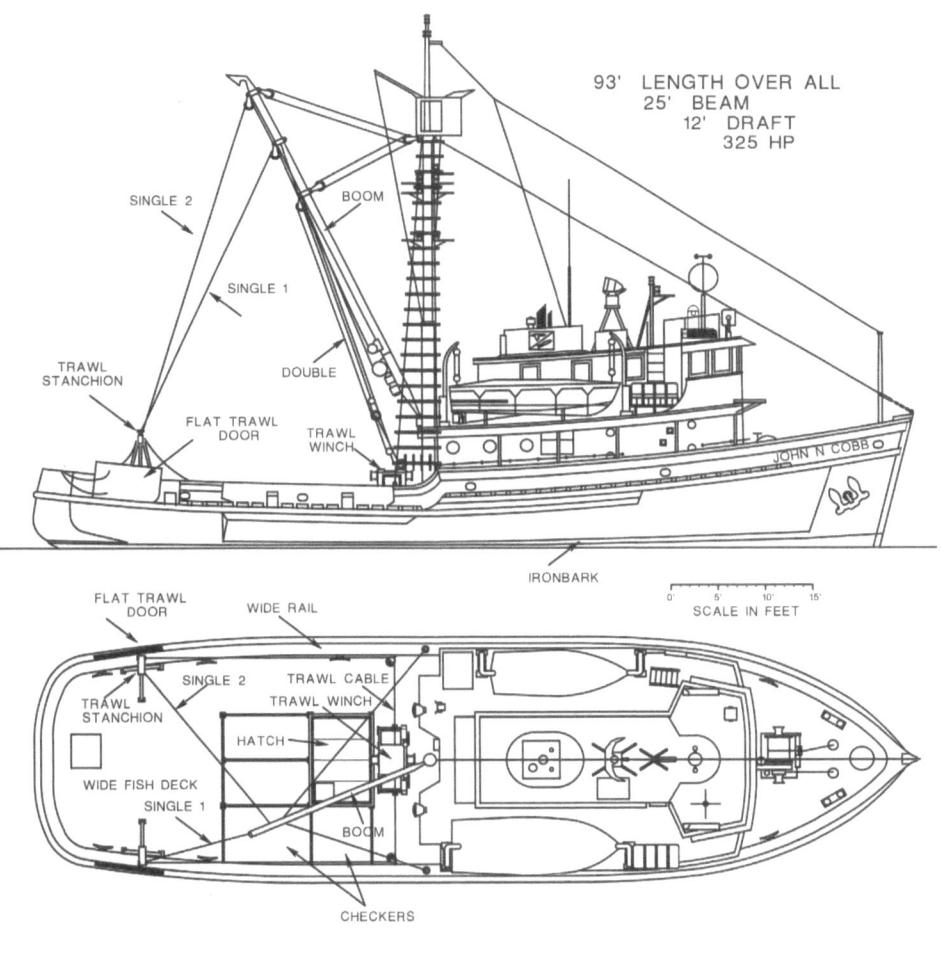

Drawing 1 R/V *John N. Cobb*—Profile (1960).

The galley, as in all vessels, is the center for the crew (Drawing 2). It's where meals are served and where the crew can relax when off watch and gather for meetings and "coffee." Coming in from the deck, a large table stood to the right, a sink on the starboard side of the galley, and an oil stove, or range, on the forward side. Directly forward was a passageway that ended with a steep stairway up to the pilot house, to the left were two heads (bathrooms) and a door to the outside, and to the right was a shower, two scientists' staterooms, the cook's pantry, and steep steps down to the fo'c'sle. There was a watertight door in the steel bulkhead dividing the fo'c'sle and engine room, giving internal access to all the ship's compartments.

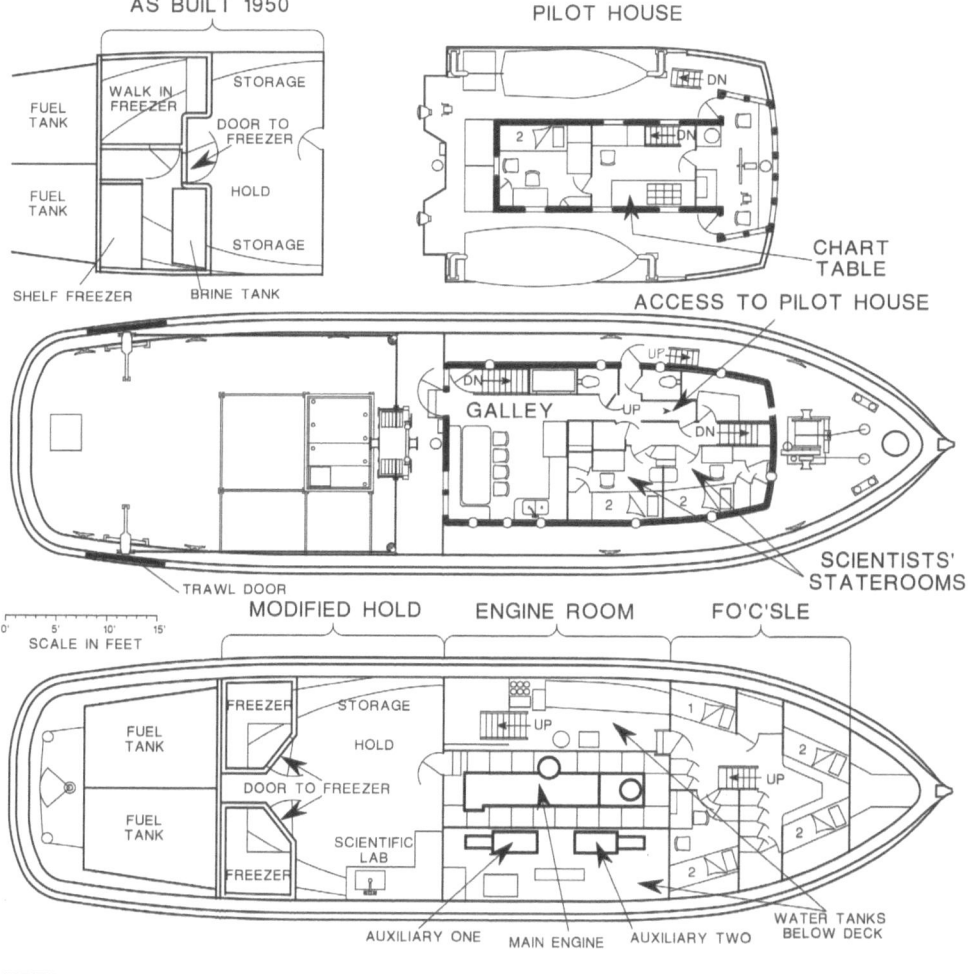

Drawing 2 R/V *John N. Cobb*—Deck Layout (1960).

Since the *Cobb* did not have to carry a large quantity of fish, as she was designed for research and not commercial fishing, she could carry more fresh water and fuel, two important items for a research vessel built in the 1950s. There were fuel tanks aft of the freezers and smaller tanks forward of the engine room under the engineer's stateroom. There were two freshwater tanks, one on each side of the main engine, and the tops of the tanks formed the engine room deck.

Fig. 4 R/V *John N. Cobb*—Pilot House (1950). Image #20047.

Aft of the pilothouse (Fig. 4) was an important item, a chart table with a flat surface to lay out a chart and drawers below for storage (Fig. 5). It was used for navigation and plotting the locations of different items. A table this size on a commercial fishing boat meant losing space for a set of bunks that could accommodate two men. With the manpower needed to harvest fish and the space needed to carry as much fish as possible, one could see why many fishing vessels excluded the table. The computer has made the chart table obsolete nowadays, and everything is recorded digitally.

Aft of the chart table was a loran set, and on the other side was a desk with a shortwave radio. The stairs from below entered aft of the pilothouse. The new Simrad sounder was mounted in the pilothouse with a modern radar set. I was in the area often, recording the location of the haul or station that the skipper or mate plotted on the chart, along with sea temperature, length of tow, etc.

Fig. 5 R/V *John N. Cobb*—Chart Table (1950). Image #20054.

The original building specs called for three different types of refrigeration: walk-in freezer, a shelf freezer, and brine tank freezer (Drawing 2 insert). They were constructed in the hold forward of the after fuel tanks. There was a central door in the freezer that opened into the hold. Once inside the freezer, looking aft to the right, was a second door that led into a large walk-in freezer. To the left was an open workspace between a forward brine tank and an aft shelf freezer, both used for experiments on methods of preserving different species of fish. By the time I inspected the vessel in July 1960, there had been modifications done to the fish hole because of dry rot, so redesigning the refrigeration space became paramount. Two separate walk-in freezers were installed, one on the port side and the other to starboard, with a separate entrance to each. The brine tank and shelf freezer were removed because they had seldom been used since she was commissioned in 1950. A scientific lab, not called for in the building specs but found to be necessary, was also built in the hold on the starboard side.

The scientific stateroom became part of my life, and once I became the Chief Scientist (CS), the desk was more important. It was the place where I could pull information together and make sure forms were filled out properly and notebooks for the trip were kept up. It was also a relatively quiet place, where I could keep my personal log. It would be a while until I became a CS. I was a scientist in training, but once I became one, Al Pruter had me and any other CS keep a personal log of their activities during a cruise, which he reviewed after each trip. They became very important to me in later years when I began to pull the information together about Exploratory Fishing and the *Cobb*.

I, along with other biologists who worked on the *Cobb,* benefited from the work of the team that designed the research vessel for Seattle's exploratory base. The design was remarkable, perfect for what we needed to do. When I inspected the *Cobb* in 1960 and went down into the engine room for the first time, I was astonished! It was spotless. After almost nine and a half years since her commissioning, the engine room appeared just like it did at her commissioning (Fig. 6).

Fig. 6 R/V *John N. Cobb*—Engine Room (1950). Image #20058.

Creation of the *Cobb* (1941–1950)

It all started December 7, 1941, when President Roosevelt declared war with Japan after they attacked Pearl Harbor. I remember it well. I was just eight years old, and we were coming home after church in Bellingham, Washington. We were just going in the front door when a neighbor came running across the street calling, "Turn on your radio!" My father went across our living room and turned on the switch of the upright radio. After the vacuum tubes warmed up, we heard an excited voice announcing that Pearl Harbor was under attack. The effect on the US was profound.

I remember the worry in our neighborhood after the war broke out, the blackout curtains that were hung in our house at night. The bomb shelter that we hurriedly put together in our basement. There was a crawl space from the basement to a space under the front door concrete porch, where we made our emergency shelter with blankets, flashlights, and a first aid kit. There was rationing of gasoline and rubber tires and shortages of butter, bacon, and beef, products that were needed by the armed services. There were paper drives that filled our grade-school gymnasium to the rafters with newspapers. As a student, I remember going from our car carrying in newspaper bundles tied with string and piling them onto stacks. There was a huge pile of aluminum pots and pans stacked on the school campus, and I remember walking up to one and throwing some more pans onto the pile. The gas rationing stickers displayed on car windshields indicated the amount of gas one could purchase per month. Blue stars displayed in front windows of neighbors' homes showed family members serving in the military. If the stars' color changed to gold, the family had been notified a member of their household had been killed in action. During the 1930s, before World War II, the United States bought canned king crab that had been caught in the southern Bering Sea off the Alaska Peninsula from the Japanese. The American fishermen wanted to know where and how to catch king crab in Alaska, so the United States Congress passed an appropriation in 1940 to investigate the possibility of establishing an American king crab canning industry in Alaska. Since the cost of conducting exploratory work would be too great for private industry, Congress authorized the Fish and Wildlife Service to make the study. It was conducted out of the US Fish and Wildlife Montlake Lab's Technological Laboratory, which at the time was one of two groups—the other was the Biological Laboratory. The Technological Lab's objective was to help industry utilize additional fisheries resources.

Money became available on July 1, 1940, and the cannery vessel *Tondeleyo* and three fishing vessels—*Dorothy, Locks*, and *Champion*—were chartered to conduct the first exploratory cruises in Alaskan waters. Because slow funding delayed sailing the vessels in 1940, only the Gulf of Alaska was surveyed. In 1941, the group completed the study by surveying the Bering Sea before December 7, when Japan attacked Pearl Harbor and the United States went to war.

The *Report of the Alaska Crab Investigation* was published in May 1942.[2] It was read and analyzed by an officer in the Department of the Interior, who recommended to the War Food Administration that the untapped commercial resource of Alaskan king crab and bottomfish should be utilized as an additional source of food. The idea was developed and became part of the national defense plan. They planned to convert an oceangoing vessel to a factory ship and construct five fishing vessels to accompany it. They would harvest the resource and deliver it to the factory ship for processing and preserving it as food for the armed services. The 410-foot WWI steamship *West Calumb*, built in 1919 and later renamed *Mormacrey*, began undergoing modification into a fishing factory ship.[3] On August 6, 1945, the atomic bomb was dropped on Japan, and the war ended much faster than anticipated.

Pacific Explorer (1945-1948)

I was twelve years old when Japan signed the surrender papers on the deck of the *USS Missouri* in Tokyo Bay on September 2, 1945. America celebrated the end of the war with V-J Day (Victory over Japan Day). I, and the rest of the general population, had no idea about the plans to convert a ship to process king crab and bottomfish from the Bering Sea for another source of food for the country. During WWII, there was a food shortage, and the nation was frantically looking for additional food resources to feed the military and the public. By the time the war ended, the modification of the *Mormacrey* had already started and two million dollars had been spent. Whether to scrap or continue became an issue. Finally, the decision was made to complete the project and turn it over to the fishing industry.

The US government would continue the plan with the fishing industry's involvement. The government's Reconstruction Finance Corporation (RFC) controlled the funding. The Pacific Exploration Co. (PEC) of Seattle was formed and headed by Nick Bez, a Seattle businessman, who would control the project and had the funding for operating the vessels. Once PEC was formed, an additional item was added to the plan. During the winter months, when the Bering Sea was too rough to fish, the vessels would head south to explore the Southern Ocean to harvest tuna.

Fig. 7 *Pacific Explorer* at Bellingham, WA. Image #1996.10.1114.

The naval architects of W. C. Nickum and Sons did the design work for the *Mormacrey's* conversion to a factory ship.[4] The vessel was moved to Bellingham, Washington, for the final stages of work, which were completed in December 1946. The factory ship was renamed *Pacific Explorer* (Fig. 7) with PEC responsible for its operation. It was taken out to sea that month for a shakedown cruise off the Washington coast. Four fishing vessels designed by H. C. Hanson, a Seattle naval architect, were built.[5] They were one-hundred-foot steel-hulled clipper-trawlers, which are better known now as Pacific coast combination vessels. The first two, *Oregon* (Fig. 8) then *Washington*, were built in Astoria, Oregon, during 1946. *Alaska* and *California* were built the following year in Long Beach, California.

Pacific Explorer sailed for her first trip on January 3, 1947, to the tuna waters off Central and South America.[6] The crew for this cruise was smaller than those needed in the Bering Sea fisheries, since the preservation of tuna was much simpler and took less manpower than processing king crab and bottomfish. Nick Bez made arrangements for a group of ten commercial seiners and bait boats to accompany her. The tuna industry had a fleet of tuna clippers that worked the waters off South America. They used the

Fig. 8 R/V *Oregon*—Rigged as a Bait Boat. Image #1982.92.8640.

same fishing technique as the bait boat—that is, to chum the tuna with live bait stored in the tanks on their stern. When the tuna were located, they were chummed into a feeding frenzy. Fishermen would stand in racks arranged around the stern and catch tuna with hand polos and un-baited hooks. There was a need to replace the live bait as it was used up. The vessel would have to catch the bait with small nets and replenish them in their live tanks aboard or purchase them locally. The clippers had to have large brine refrigeration tanks to preserve the tuna until they arrived at a shore plant back in Southern California for final canning. The clippers were also designed for speed and were beautiful as they transited from Southern California to the fishing grounds off South America and back. The *Pacific Explorer* would supply the small fleet of vessels that Bez had put together with fuel and water and take the tuna they caught, preserving them by freezing for future canning. If the small seiners caught enough tuna without live bait, they would not have to depend on live bait. The small fleet preceded south and began fishing. The newly built *Oregon*, rigged as a bait boat, was added to the mix. Most likely, the *Pacific Explorer* anchored in protected waters outside the three-mile limit in the Gulf of Nicoya, Costa Rica, and began receiving tuna and freezing them for future processing back in Oregon.

After they had been in Costa Rica for five months, the contract between the PEC and the US government was suddenly cancelled. The order was issued, recalling the two US-financed vessels, *Pacific Explorer* and *Oregon*, to the United States.[7] The *Pacific Explorer* arrived in Astoria, Oregon, on July 23, 1947, with 2,250 tons of tuna aboard. The *Oregon* arrived about the same time.

When the Pacific Fishing Project was proposed during the war, it was fine, but after the war, it was against the law. Southern California commercial fishery and canning industries complained that the PEC allowed a vessel backed by US tax dollars to be in direct competition with private enterprise. The project needed to be more scientific than commercial, so the contract was rewritten, transferring the *Pacific Explorer* and the four fishing vessels from PEC to the federally owned RFC. Operation of the vessels would still be conducted by the PEC. A new plan was drawn up, which explains why, after the *Oregon* returned to Astoria, none of the four fishing vessels ever landed their catch on the *Pacific Explorer*. Instead, they were used to explore different areas. The *Alaska* conducted an exploratory trawling trip to the Bering Sea in July 1947.[8] Then the *Alaska* was converted to a tuna seiner and the *Oregon* as a bait boat and were used in 1948 to explore the Western Pacific where the Japanese had fished before the war.[9]

The *Pacific Explorer* sailed on March 26, 1948, for her second trip, heading for the Bering Sea to fulfill the objective of the World War II Pacific Fishing Project for which she had been modified.[10] She was to harvest the fisheries resources believed to be in the Bering Sea, namely king crab and flatfish. Nine commercial fishing vessels were chartered by the PEC to supply the factory ship with the catch that was to be processed on board. The four fishing vessels built as catcher vessels for the mother ship were not used. They found commercial quantities of king crab and flatfish in the Bering Sea. From the catches, they canned part of the king crab and froze the rest. The flatfish went through the filleting line before the fillets were frozen, while the rest of the commercial fish were frozen whole. The rest of the catch and the fish waste from the filleting line were converted into fishmeal. By July 5, 1948, all the fishing vessels had fulfilled the ninety days of fishing their contracts called for and had departed. The *Pacific Explorer* returned to Astoria on July 18, 1948, to offload their catch.

I looked at old pictures of the *Oregon* and *Washington* in the office archives and listened to the rumors of why the *Pacific Explorer* failed and wondered how much money the fishermen lost on the project. When I was at a luncheon for retired biologists, I described the *Pacific Explorer's* history to a few interested people, saying it would be fascinating to interview someone who had been on the Bering Sea trip, but how would we find someone after sixty-six years? Ed Best, a retired biologist who was listening, said, "I was there as a fisherman on the *Kiska*." I believed that Nick Bez made a lot of money off this adventure and left the fishermen without getting paid, so I asked, "Were you paid?" because the adventure was a financial disaster. Ed replied, "We were under contract and the pay was good." He had retired from the International Pacific Halibut Commission, and our paths had crossed over the years, so I asked him if we could get together and talk about his experience.

Later, he told me of his adventures in 1948, when he had just returned from World War II after serving with the Seventy-First Infantry Division in Europe. He was raised in Gig Harbor, Washington, a fishing community where his stepfather owned a salmon seiner, and Ed had been part of the crew since he was thirteen. One evening after his return from the service, Ed was going into a bar when a bearded fellow came out and asked, "How would you like to go to the Bering Sea this summer and make some money?" Ed said yes, and that was the start of the adventure.

He went to Astoria, Oregon, where the *Kiska* was located, and worked as part of the crew, getting her ready for the trip. There were six in the crew—the skipper, engineer, cook, and three fishermen. The vessel was owned and operated by the Columbia River Packers Association, with which Nick Bez was affiliated. The *Pacific Explorer* had recently landed tuna from her first trip to South America, where the *Kiska* had worked as a bait boat. She was the only one chartered for both *Pacific Explorer* trips. The top two boats were *Sunbeam* and *Kiska*, who exceeded the quota before the ninety days were up. They were able to continue fishing, landing fish and king crab, and were paid extra at the daily rate the contract called for (Fig. 9). By July 5, 1948, all the fishing vessels had fulfilled their

Fig. 9 *Pacific Explorer* with the *Tordenskjold* and *Seiner* alongside. Image: Ed Best Picture.

contracts and had departed. This interview with Ed sure changed my mind about the project.

Results from the two trips indicated that the *Pacific Explorer* could have supplied needed protein to the troops during the war. During peacetime, there was a need to make a profit, but because they did not, the operation was shut down, ending the Pacific Fishing Project. The project did point out, however, that there might well be more unknown resources along our coastal waters and that we should find out if they existed. If we ever went to war again and needed additional food, we would know where to get it—that was what started a new branch of research. In 1949, the US Fish and Wildlife Service added a third group, Exploratory Fishing, to the other two, Biological and Technological. They also received two of the four fishing vessels from the project as exploratory research vessels. The first vessel, *Washington,* was transferred to Seattle in 1948, and the second, *Oregon,* was transferred to Pascagoula, Mississippi, in 1949.

Birth of the *Cobb* (1948-1949)

On July 1, 1948, Congress authorized funds for the operation of an exploratory fishing vessel in Alaskan coastal waters. But until the funds became available, nothing could be done, and the time to be in the Bering Sea was passing with the coming of fall. The northern part of the Bering Sea had not been explored, whereas the southeastern portion had been. King crab and bottomfish were found in commercial quantities there. How far west and north did they extend?

When money was finally allocated, fishing gear was obtained for the *Washington,* a crew hired, and stores put aboard. Even with late sailing, the office decided to send the untried vessel into the Bering Sea to see how it would operate under notoriously bad weather conditions, which should be equal to the worst the vessel would likely encounter in the future. The *Washington* (Fig. 10) finally departed Seattle on August 24, 1948, en route to the Bering Sea for the first exploratory cruise conducted by the newly formed exploratory base.[11]

Once in the Bering Sea and heading north, she ran into severe weather. It was discovered she was not ballasted correctly; there was not enough weight in the stern. The skipper was unable to keep her bow pointed into the oncoming storm waves without following off or broaching. I assume the skipper had to turn around and run with the waves, losing many miles

Fig. 10 R/V *Washington*—Before Departure. Image #11943.

that he would need to make up once the weather moderated. Fortunately, once she reached Nome, Alaska, she received thirty tons of ballast in the form of rock loaded in her aft cargo hole, which helped solve the problem. The *Washington* began survey work on September 14, 1948, in the vicinity of Nome and returned to Seattle on October 23, 1948. Each of the scientists made individual reports, basically recommending that the *Washington* needed to go through major modifications in order to make it a usable exploratory vessel (Drawing 3).

Her good points were mentioned first. She was designed as a clipper-trawler, or a large combination fishing vessel, which, with minor modifications, could be converted to various commercial fisheries, ideal for an exploratory fishing vessel. There is no question that the east coast side trawler design was a better sea boat, as demonstrated by the *Deep Sea,* which was working the Bering Sea at the time, but she was good for one fishery only—trawling. The *Washington's* skipper concluded that she had a strong and seaworthy hull, her mechanical condition was good with adequate power, and she was self-contained with reserves of fuel oil, lubricating oil, dry and frozen provisions, and fresh water for long trips. She had good refrigeration facilities, which were designed for a commercial operation.

Suggested modifications followed. A higher freeboard was needed aft of the house, since the deck was awash in any kind of weather. If the weather got worse, waves could come over the rail, washing the catch back

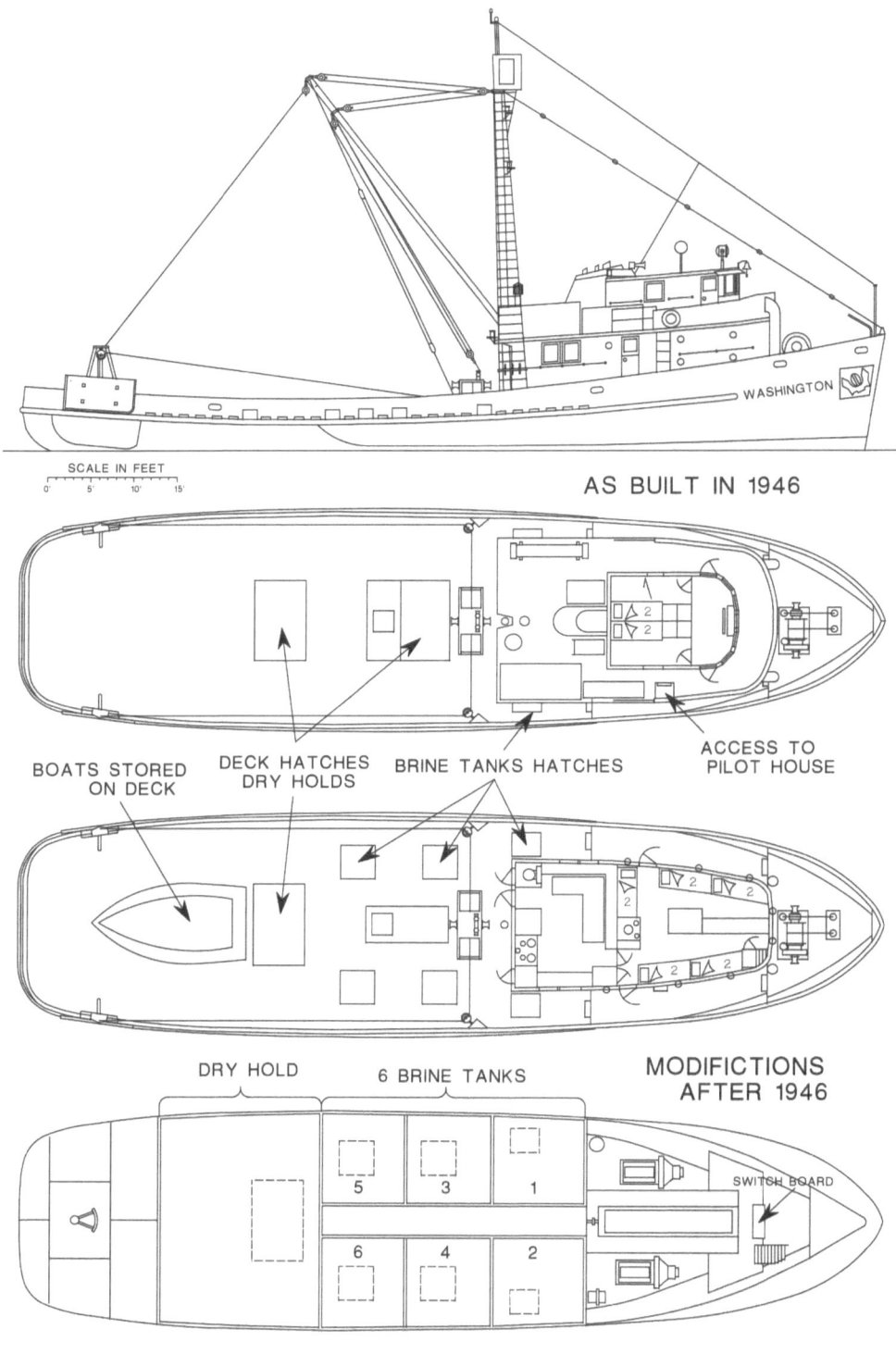

Drawing 3 R/V *Washington's* Modifications.

into the sea. Scientific bottled samples were lashed down on the afterdeck and were also lost due to severe weather. Since the vessel was designed to carry tons of fish, the correct ballasting would have to be determined and added to the vessel. Dedicated space with easy access from the deck, space for extra fishing gear and supplies, scientific supplies and storage for dry samples, and freezer space for biological samples were indicated, along with a workboat and a skiff stored on the upper deck. On the return trip, while crossing the Gulf of Alaska, the boats stored on the stern came loose during severe weather and were damaged. Leaks in the engine room, especially one over the switchboard, and those into the crew's quarters and galley needed to be stopped. The small and crowded pilothouse needed to be replaced. Access to the pilothouse needed to be internal, not external. And finally, a general overhaul replacing weak or unsuitable equipment, such as the automatic steering mechanism and the pilothouse engine room controls, was suggested. If PEC hadn't made the radical changes to the *Washington* after she was built by adding four brine tanks (3, 4, 5, and 6 as shown in Drawing 3), she may have been accepted as an exploratory vessel as was the *Oregon*. Tanks one and two were built in all four of the fishing vessels when they were delivered from the shipyard.

Tom Dunatov, a good friend who was the first mate and skipper on the *John N. Cobb* from 1972 to 1990, fished on a tuna vessel out of California when he got out of the navy in 1957. He told me that during that time, brine tanks were commonly used in the tuna industry to carry extra fuel on long trips to the fishing grounds off Central America. Once they arrived, the fuel that was used on the trip down would have emptied the brine tanks, which were then cleaned out with detergent and refilled with sea water. The temperature would be brought down to freezing by refrigerated coils in the tank. As fresh fish were caught, they would be placed into the tank of freezing water. Adding more salt to the mix would bring the temperature down even further, so the fish would be completely frozen. When the tank was full of frozen fish, the brine would be pumped overboard and the fish moved to the refrigerated hold where they would be stacked and held frozen. This increased the holding capacity of the vessel by reducing the space between fish. The process repeated until the vessel was packed full of frozen tuna.

Before the *Pacific Explorer* had returned from the Bering Sea trip, the PEC made plans to send two of their fish boats, the *Washington* and *California,* to South America to scout for tuna to see if it would be worthwhile to send the *Pacific Explorer* to the far south for her third trip. The four brine tanks were added to the *Washington* to carry extra fuel for both vessels on their long trip and to take frozen tuna they had caught during the survey back. The *Washington* and the *California* were crewed up and ready to go and were scheduled to sail in 1948 to explore the waters off the coasts of Peru and Chili but were cancelled at the last moment due to the collapse of the project.

The US Fish and Wildlife Service Washington, DC, office made the final decision not to modify the *Washington* but to sell her and use that money to build a new exploratory research vessel. She sold for $150,000. I had always believed that the *Washington* was one of those vessels moved to China through the United Nations Relief and Rehabilitation Administration, since there were funds available in this group for suppling fishing vessels after WWII. China was undergoing an acute food shortage after the war, and the plan was to supply fishing boats and a training program to teach the Chinese how to fish. I found in a 1949 issue of the *Fishermen's News* that the *Washington* was transferred to the European Cooperation Administration for use in Korea,[12] which apparently was where the money came from to build the *Cobb*.

At least four convoys of fishing boats sailed from the Pacific Northwest coast of the United States in 1946 on the southern route to Shanghai via Seattle, stopping at Honolulu, Wake Island, and Guam. There were seven fishing vessels per convoy, each manned by a crew of seven. Tom's dad, Joe Dunatov, was one of the fishermen on the *E. E. Johnson,* part of one of those convoys to China. He remained there for a short time as a teacher because of his extended knowledge of making and mending nets. He was the lead fishermen, or bos'un, on the *Cobb* from 1956 to 1968. I got to know him when I started going out on the *Cobb* in 1960 and wish I had asked him about the *Washington*. Tom said years later that his dad was part of the crew that delivered the *Washington,* he thought, to the Philippines or Samoa.

A set of specifications was drawn up by the Fish and Wildlife Service. They were put out for bid, and W. C. Nickum and Sons of Seattle was selected as the design team. They came up with the specs for a dedicated Exploratory Fishing and Gear Research vessel, which was the first step toward building a vessel for the Montlake Lab's Exploratory Base.

Building the *Cobb* (1949–1950)

The new design for the *John N. Cobb* was approved. The company selected to build her was Western Boat Building Co., Tacoma, Washington. Through the years, I've wondered about the Western Boat Building Company. Although I was aware of the companies that built the modern tuna seiners and crab boats coming out of the area during the '60s, I hadn't paid much attention to the shipyards of Tacoma. I do remember that the mate of the *Cobb* called my attention to a vessel overtaking us on one calm evening as we headed out of the Strait of Juan de Fuca to start another cruise. She ran smoothly through the water and must have been making 12 knots or more to our 9.5. She was over twice the size of the *Cobb* and had beautiful lines that gave her the appearance of a millionaire's yacht. She cut the water like a knife and slid through it so smoothly; she was built for speed. As she drew near, the setting sun reflected off her newly painted hull and superstructure. The large mast and boom gave her away as a newly constructed commercial tuna seiner as she passed us. The California clipper fleet was going through a major change in converting from pole and line clippers to tuna purse seiners.[13] The clippers were large vessels, and it was believed that they could not be converted to purse seiners for tuna. They could and did and now were being constructed in the shipyard, replacing the tuna clippers. This was my first observation of one of these beautiful vessels. She was on her way from a Tacoma shipyard that had just commissioned her to her new home port in Southern California. The mate, who lived in Tacoma, told me the city had a key role in making tuna and sardine seiners, and now the newly constructed large purse seiners, for the fishing industry and was one of the world leaders in fishing vessel construction.

When the *John N. Cobb* was decommissioned at the NOAA facilities at Sand Point in Seattle on August 13, 2008, after more than fifty years of service to the country, there were at least two rows of people at the ceremony who had come up from Tacoma. One of the men got up and introduced himself as Allen Petrich, whose grandfather had built the *Cobb*. He noted that the people with him were either related to him or had worked for years at the Western Boat Building Company. I had the opportunity to talk with Allen after the ceremony. He remembered the vessel well, and one of his cousins said he remembered playing in the shavings below the boat while it was being built. Allen was gracious enough to fill me in on the history of the shipyard. His grandfather, Martin A. Petrich, along with

Joe Martinac, founded the Western Boat Building Company in 1916, and it became one of a number of companies building commercial fishing vessels in Tacoma.

In 1950, there were four major shipyards producing them: Western Boat Building Co., J. M. Martinac, Tacoma Boat, and Martinolich Shipbuilding. They all produced fishing vessels, herring and sardine seiners, and tuna clippers during the years when wood was preferred for construction, as well as in the '60s when steel became the material of choice. J. M. Martinac and Tacoma Boat became famous for their tuna seiners and Martinolich for king crabbers. Western Boat Building Co. got out of building commercial fishing vessels after a fire destroyed the yard in August 1950, six months after the *Cobb* was launched. It had been the largest enclosed shipyard where a vessel 250 feet long could be built on the Pacific coast. In 1948, the *Mary E. Petrich* was launched from this yard. The 150-foot tuna clipper was the largest and fastest one built at that time. After the fire, the company took a different direction, building pleasure yachts with the brand name Western Fairliner.

Like many shipyards, Western Boat gave each vessel built a hull number—the *John N. Cobb's* number was thought to be lost in the fire. The contract for the *Cobb* called for photographs to be taken during the building of the vessel. Kenneth G. Ollar of Tacoma was the photographer. In our old files, I found several black and white 8.5" x 15" prints of the *Cobb* with a stamp of the date the picture was taken and the Western Boat hull number "192" on the back of each. One of the pictures was of the keel laying on August 10, 1949 (Fig. 11), the second was when the house was starting to take shape on November 21 (Fig. 12), and the third was taken when she was launched on January 16, 1950, sliding into the city waterway below Tacoma's Eleventh Street Bridge (Fig. 13). The US Fish and Wildlife Service had its own numbering system for their vessels, and the *Cobb's* Official Number was F.W.S. 1601. The vessel was accepted by the US government on February 13, 1950, and she was commissioned at the University of Washington Oceanography Dock in Portage Bay, Seattle, on February 18, 1950. It took about five months to build the *Cobb*, from laying the keel to launching.

The US Fish and Wildlife Service's Seattle Exploratory Fishing and Gear Research base finally had its dedicated research vessel. The new vessel *John N. Cobb* sailed on her first shakedown voyage on March 24, 1950, which was the fourth exploratory cruise conducted by the base.[14] She had been operating for ten years by the time I was hired in July 1960. She was dedicated to the base and had completed her forty-sixth exploratory cruise; by that time, they were referred to as the *Cobb's* cruises.

Fig. 11 R/V *John N. Cobb*—Keel Laid. Image #20002.

Fig. 12 R/V *John N. Cobb*—House and Deck Work. Image #20014.

Fig. 13 R/V *John N. Cobb*—Launched. Image #20036.

Chapter 1: Endnotes

1. J. G. Ellson and Sheldon W. Johnson, "The Exploratory Fishing Vessel John N. Cobb," *Fishery Leaflet* 385 (October 1950). MS #4 and #5.

2. "Report of the Alaska Crab Investigation," in *Fisheries Market News* Vol. 4, no. 5a, May 1942—Supplement, pp. 1–108.

3. Carl B. Carlson, "*S.S. Pacific Explorer*—A Preliminary Description," *Commercial Fisheries Review* 9, no. 1 (January 1947). Also Sep. No. 161.

4. George C. Nickum, "Pacific Coast Processing Vessels," in *Fishing Boats of the World*, ed. Jan-Olof Truang, (London, England: Arthur J. Heighway Publications LTD., 1955), pp. 513–534.

5. H. C. Hanson, "Pacific Combination Fishing Vessels," in *Fishing Boats of the World*, pp. 187–202.

6. Carl B. Carlson, "*S.S Pacific Exporer*—Part III—Below Deck Arrangements And Refrigeration Equipment," *Fishery Leaflet* 316 (August 1948).

7. Carmel Finley, *All the Boats on the Ocean* (Chicago: University of Chicago Press, 2017), p 64, pp 211.

8. Joseph E. King, "Experimental Fishing Trip to Bering Sea," *Fishery Leaflet* 330 (March 1949).

9. O. R. Smith and M. B. Schaefer, "Fishery Exploration in the Western Pacific (January to June 1948 by Vessels of the Pacific Exploration Company)," *Commercial Fisheries Review* 11, no. 3 (1949).

10. Norman B. Wigutoff and Carl B. Carlson, "*S.S. Pacific Explorer*: Part V—1948 Operations in the North Pacific and Bering Sea," *Fishery Leaflet* 361 (January 1950).

11. J. G. Ellson, Boris Knake, and John Dassow, "Report of Alaska Exploratory Fishing Expedition Fall of 1948, To Northern Bering Sea," *Fishery Leaflet* 342 (June 1949), MS #1.

12. *Fisherman's News* 5, no. 9 (September 1949), pp. 3. Photograph of the *Washington*.

13. Richard L. McNeely, "Purse Seine Revolution in Tuna Fishing," *Pacific Fisherman* 59 (7) (June 1961), pp. 27–58. MS #63.

14. Edward A. Schaefers, "The *John N. Cobb's* Shellfish Explorations in Certain Southeastern Alaska Waters, Spring and Fall of 1950 (A Preliminary Report)," *Commercial Fisheries Review* 13, no. 4 (April 1951), MS #7. See also Sep. No. 278.

CHAPTER TWO

Lee Alverson's Interest 1949–1959

In 1949, when I was in high school, my future boss, Dayton L. Alverson, had a summer job tagging sole for Oregon State on a commercial dragger while he was at the UW College of Fisheries. His job was to contact various trawlers and ask the skipper if he could go out on the vessel and tag some of the flatfish they had caught. One of the skippers he went out with who ran the commercial dragger *Harold A.* describes a surprising catch in Lee's book *Race to the Sea*.[1] The skipper said, "Well, I don't know how many Petrale or English sole we are likely to catch because I'm going to be fishing rather deep, but there will damn sure be a lot of 'shit sole' (Dover sole or slime sole) to be tagged." Lee went along, and they made a two-hour tow with an otter trawl that skimmed the bottom at 90 fathoms (540 feet). The net came up with ten thousand pounds of rockfish. Lee couldn't believe the diversity of rockfish species. He used a fish key entitled Fishes of the Pacific coast of Canada[2] to determine the species that made up the bulk of the catch. He found out that it was the longjaw rockfish (*Sebastes alutus*). I believe this was the spark that stimulated Lee's interest in this resource that Oregon fishermen referred to as "rosies."

Lee Alverson was close to graduating from the UW College of Fisheries in 1949; he lacked three credits.[3] Don Powell, a classmate, had just been hired by the new Exploratory Fishing and Gear Research Base. He talked Lee into applying for a job with this fledging group. At the end of the summer of 1949, Lee returned to Seattle from the part-time job he had in

Oregon and applied for a permanent one at the new Exploratory Fishing and Gear Research Base. He enrolled in the fall quarter, working on the rockfish and developing a key for their identification, which was published.[4]

Pacific Ocean Perch — How It Got Its Name

In 1854, Dr. Ayres described the first three rockfish taken in California waters—the bocaccio, China rockfish, and the yelloweye rockfish—and placed them into the genus *Sebastes*. There was considerable confusion as more rockfish species were found and placed into different genera. Finally, in 1895, Frank Cramer determined that the rockfish would be placed into one genus, *Sebastodes,* and from 1949 to 1971, the majority of rockfish species were known as *Sebastodes*. In 1971, scientists reached a universal accepted conclusion that they all should be in the original genus, *Sebastes*.[5] In this book, from this point, for the generic name of any species of rockfish referred to as *Sebastodes,* a capital "S." will be used followed by their scientific species name. Because the time period of the story in this book is from 1949 to 1970 and *Sebastodes* was used in all the reports and literature generated during that period. The genera of the rockfish in this book will be referred to as *Sebastes* to keep the reader familiar with the established official genera name now.

Fish names are interesting. There are two, the scientific name and the common name. Scientific names are based on a biological classification system and are made up of two names: a generic and a species. The generic name applies to animals with similar characteristics. The species name is a group of individuals that interbreed among themselves. For example, *S. alutus* live in the northeastern Pacific with at least sixty-three other rockfish species of the same genus, with which they do not breed. Common names are used locally by local populations referring to individual fish, but there is also an official common name. I have followed the list that is in Love's 2002 book.[6] The twenty-four rockfish of the genus *Sebastes* that I was involved with and are discussed in this book are listed alphabetically by their scientific name followed by their common names in Appendix 1.

A third name exists, *S. alutus,* known in the market as ocean perch, which is misleading since perch generally refers to a freshwater bony fish, whereas *S. alutus* is found in salt water. I visited Jergen Westrheim,[7] an expert on rockfish and Lee Alverson's classmate, on June 2, 2010, at his home in Nanaimo, Canada. During our visit, he told me a story about how the ocean perch got its name. In the '50s, when he began working for Oregon State, the groundfish fisheries was new, and commercial fishermen were trying new ideas to expand their market, like deepwater trawling along the continental break of 100 fathoms. They caught a species of rockfish called the longjaw rockfish (*S. alutus)* or, as the Oregon fishermen called them, "rosies." But because there was no market for it, they brought a load into Newport and gave it to the fish house to see what could be developed. They found a market on the East Coast, where an East Coast species of redfish, *Sebastes marinus,* was being sold as ocean perch. When two different species live in different oceans, but their fillets look alike, their texture and taste are similar, and they are shipped to the same location and sold as ocean perch, it becomes a problem.

Jergen learned that marketing for this species started with the Great Lakes' yellow perch (*Perca flavescens*) fishery. It produced a fillet with the skin on, which sold well in the Midwest, especially during the war years, but the yellow perch catch declined even while there was still a great demand for it. On the Atlantic coast, there was a species of rockfish, the rose fish, *Sebastes marinus,* which was filleted like the yellow perch, with the skin on, and sold in the yellow perch market as perch. The customers would say, "The product is excellent, but why is the skin red?" The answer, "It's from the ocean," resulted in the name "ocean perch." It had better flavor and texture than yellow perch and was available, so it became popular. The Pacific coast longjaw rockfish, *S. alutus,* began selling in the same market under the same name.

Jergen told me that two refrigerated boxcars arrived in Denver at the same time, both loaded with fish fillets, carrying the name "ocean perch," one from the East Coast and the other from the West Coast. The one from the West concerned the East Coast vendors since the fillets were larger and apparently more desirable. East Coast vendors became upset and brought a case to court saying you can't use that name since they are two different species and, in fact, are two different genera, *Sebastes marinus* vs. *Sebastodes alutus.* They sued the West Coast supplier, saying they could not use the same common name for different species. The case was handled in federal court in Portland, Oregon. Experts gave testimony, including Dr. Welander, a professor at the UW College of Fisheries, and his student, Lee Alverson, who published a key for the rockfish identification. There was so much confusion regarding the scientific names, especially about which genera the species should be classed, that there was no clear-cut conclusion that they were different species. The judge finally ruled that the common name of the Pacific species, longjaw rockfish, could be changed to "Pacific ocean perch" (POP) with a small "o," and the East Coast species could retain the name "ocean perch."

I have always been confused regarding the Atlantic redfish fisheries and understood from the presentation by Clarke's in *The Encyclopedia of Marine Resources* edited by Frank E. Firth that for many years there were two species of redfish taken in the fisheries: *Sebastes marinus* and *Sebastes viviparous*.[8] I have always considered that *Sebastes marinus* was the one competing with *S. alutus,* both sold under the market name ocean perch. After the decision to place the rockfish in the genus *Sebastes* in 1971, a list of all the known *Sebastes* species was established and is published in Love's book.[9] There are four species of redfish in the Atlantic Ocean—*Sebastes fasciatus* (Acadian redfish), *Sebastes mentella* (deepwater redfish), *Sebastes norvegicus* (golden redfish), and *Sebastes viviparus* (Norway or small redfish). *Sebastes marinus* isn't even listed because it is not a valid species and the proper name should be *Sebastes norvegicus*.[10] The market name for the East Coast rockfish is still ocean perch, while for the Pacific Coast, it is Pacific ocean perch (POP). In this book, I have continued using the scientific name of *Sebastes marinus* for the East Coast species of ocean perch that is in competition with the Pacific ocean perch, which probably should be *Sebastes norvegicus,* the golden redfish.

First Deepwater Exploratory Cruises

Lee now had a full-time job in 1950 and could start looking for new resources in the waters on the northeastern Pacific Coast. I am sure that the experience of observing the huge catch of rockfish made by the Oregon dragger *Harold A.* stimulated his interest in exploring deep water for rockfish along the top of the continental slope. How deep the POP went and what other species of rockfish mixed with them at the top of the continental slope were unknown at the time, which further stimulated his enthusiasm about the job.

He wanted to begin exploring the deepwater area where the continental shelf ends and the slope begins,[11] but he had to fight for vessel time on their newly built research vessel, the *John N. Cobb,* since the tuna industry had taken up all the summer months of good weather for the albacore cruises. He finally got the first trip scheduled for 1951. It would occur during the equinox when summer ends and fall starts, September 22, which is also when the fall and winter storms along the Pacific Coast are expected to appear. It wasn't the best time, but it was feasible.

Cruise 9 (1951)

It was the first deepwater trawl survey conducted by the exploratory base. The area of operations was in waters off the coast of Washington, west of Cape Flattery, sailing from Seattle on September 27 and returning on October 19, 1951. Before the cruise began, Lee and others went over the Washington coast navigation charts to decide which area to survey. They decided on the bottom depths that were 100 fathoms or deeper along the continental slope and the canyon that extended northeast toward the lightship station *Swiftsure*, splitting the continental shelf into two parts (Drawing 4).

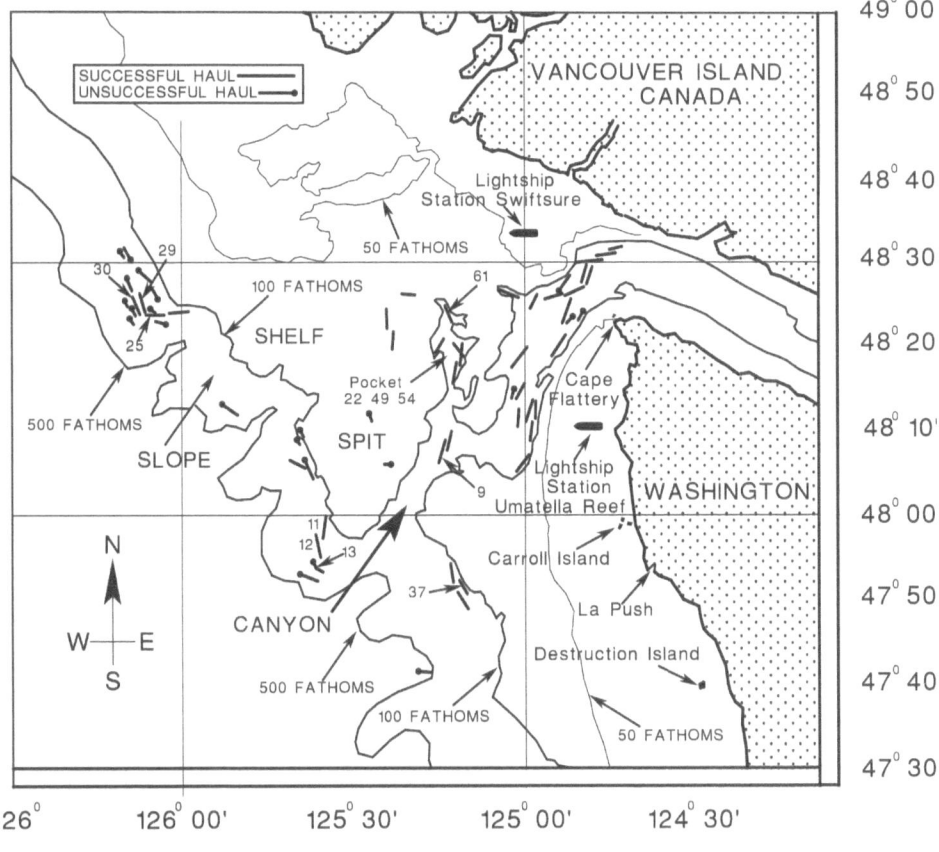

Drawing 4 R/V *John N. Cobb*—Cruise 9 Hauls.

Fishing gear was a standard commercial four-hundred-mesh trawl used by the local trawl fleet. Floats along the head rope had to withstand extreme pressure in deep water, so eight-inch diameter deep-sea aluminum alloy floats manufactured in England were used, guaranteed to withstand pressure to 530 fathoms (3,180 feet). They used the vessel's echo sounder to mark whether the bottom was flat or rough. When the bottom was found to be flat and smooth, the decision was made to try to tow the net. If the net could be towed along the bottom at approximately the same depth for an hour without snagging, hanging up, and stopping the vessel, the haul was considered a success. You can imagine the amount of time spent in sounding for a possible drag and the amount of repair work done on gear when the net snagged on the bottom or when the trawl hung up on an unknown object.

The fishing log attached to the final report gives the results of the survey, listing the actual catch of each haul in pounds.[12] It is divided into three parts: flatfish, round fish, and rockfish. The study found commercial quantities of Dover sole, sablefish, and red rockfish. One of the species of the red rockfish was the POP. A market had been found for POP along the East Coast, and there was an intense interest in them from the Pacific trawl fleet — how large was the resource, how deep did it extend, and what other rockfish were found in the deeper water? A catch of a thousand or more pounds was considered large.

Out of sixty-one hauls made during the trip, six were made on the shelf, twenty-nine in the canyon, and twenty-six on the slope. Thirty-seven were successful, but twenty-four were unsuccessful, snagged, hung up, or the doors were crossed. The latter is a serious matter. When a trawl is set, the trawl doors sheer off to the side like a kite, one to the right and the other to the left, which spread the net. Floats on the head rope and the lead line on the foot rope open the net. If for some reason the doors do not sheer to the side but turn inward and cross over, the two cables wind together and close the net. To untangle the mess, it takes experience and knowledge and is dangerous.

The twenty-four successful hauls were made in the canyon in depths of 100 to 184 fathoms. Three of them had no POP, sixteen had catches between 40 and 700 lb. of POP, and five hauls were 1,000 lb. or more of POP. Haul 9, the closest to the outside, where the canyon ended at the break of the continental slope, had 3,000 lb. of POP taken at depths of 158 fathoms. The other four were in a pocket just north of haul 9 — haul

22 took 1,000 lb. at a depth of 106 fathoms, haul 49 took 2,800 lb. at 108 fathoms, haul 54 took 1,600 lb. at 100 fathoms, and haul 61 took 1,200 lb. at 102 fathoms. There were four unsuccessful hauls made in the canyon.

The Spit area on the shelf is well known by commercial fishermen but not by the general public. It is not labeled on the navigational charts and can be confused with the Dungeness Spit that is located in the Strait of Juan de Fuca near Port Angeles. The latter is a danger to navigation since it is just above the high tide mark, so there is a lighthouse on the end of the Dungeness Spit. When the fishermen try an old net in prospecting for new trawl hauls in the Spit area off the coast of Washington, if it went through, there were often good catches of fish. Of the six hauls made on the shelf during the survey, the three made on the south end of the Spit hung up. The other three, made north from there near where haul 22 was made, were successful. All three had small catches of POP that ranged from 200 to 700 lb. One of them also had a good catch of a commercial species of Dover sole.

The surveyed bottom along the continental slope outside of the canyon, between 100 fathoms and 250 fathoms, was not as trawlable as in the canyon. Out of twenty-six drags, ten were successful, and sixteen were unsuccessful. I am sure Lee was excited with the results of the POP catches of the successful ones. Haul 12 took 3,800 lb. at a depth of 148 fathoms, and haul 11 took 1,000 lb. in 114 fathoms. Haul 13, though classified as unsuccessful by snagging, had the largest catch at 5,000 lb. of POP, and it was made in deeper water of 225 fathoms. All three were grouped together where the canyon split the continental shelf. Further north on the slope, three hauls produced good catches of POP—haul 25 took 1,200 lb. at a depth of 125 fathoms, haul 29 took 1,000 lb. at 130 fathoms, and haul 30 took 4,000 lb. at 150 fathoms. This wasn't the best area to trawl, and the other ten hauls hung up or snagged badly, losing the catch. To the south, on the east side of the canyon opening, haul 37 produced 1,200 lb. of POP at a depth of 108 fathoms.

Excluding POP, four species made up the red rockfish catch: *S. crameri*, *S. diploproa*, *S. introniger*, and *S. rosaceus*, along with an additional species, the "idiot" or shortspine thornyhead (*Sebastolobus alascanus*). The latter is not of genus *Sebastes*, a true rockfish, but was still included in the red rockfish category. They were found in the canyon and outside along the continental slope. The catches of the four *Sebastes* species varied from 100 to 750 lb. with the exception of haul 13, which had 1,100 lb. of *S. introniger*. Loves' 2002 document states that it is synonymous with *S. aleutianus*.[13]

It was the first proof of the vast resource of POP Lee had envisioned, hidden in the deep water over the top of the continental slope, one that had been hidden from humans from the beginning of time. Commercial catches of 1,000 lb. or more were found from the top of the continental slope at 100 fathoms to 225 fathoms, the deepest haul made in the study. The deepest haul produced the largest catch of POP at 5,000 lb. How much deeper did the resource extend? If this resource had been discovered during WWII, and we'd had the freezing capacity we do now, it would have helped supply the needed protein to feed our massive military appetite. Now, the military were contracting fillets from both the East and West Coasts for ocean perch to feed the troops during the Korean conflict.

I can imagine the impatience Lee had to endure after completing Cruise 9 before going out on another deepwater trawling cruise a year later. Since the group was small, he needed to participate in other exploratory group missions as well as fight for bottomfish cruise time on the *Cobb*. The tuna industry was lobbying for the *Cobb* to scout for albacore and exclude the POP exportation. Lee wrote a memo to the director of the exploratory base to stop the extension of the 1952 albacore work. In a meeting held in Seattle in November 1951, the fishermen had been led to believe that the *Cobb* would spend several months the next summer exploring the deep water along the top of the slope for POP. Lee indicated to the director that the otter-trawl fishery, which had grown from 75 offshore boats in 1951 to 181 in the first few months of 1952 because of the interest created by the *Cobb's* trawl work in 1951, had caused a large number of combination vessels to convert to otter-trawl fishing, and they expected an addional POP cruise. Lee concluded by stating, "It seems we are pulling out at a time when we are needed most. The proposed dragging trip in November is, of course, out of the question because of the weather. It would be looked on as a mere token operation at a time when exploratory dragging is impractical. Surely, we have nothing to gain by such a trip and could be subjected to a considerable amount of ridicule." Finally, he was assigned the second deepwater trawl survey. Now, he knew from Cruise 9 that they were found as deep as 225 fathoms, and the best catches occurred between 100 and 220 fathoms.

Cruise 13 (1952)

Lee took the *Cobb* out on the second deepwater trawling cruise, which departed Seattle on August 25 and returned on October 3, 1952.[14] Again, before they left, they used the nautical charts of Washington and Oregon to determine areas of 100 fathoms and deeper that they wanted to survey. The top of the slope between 100 and 500 fathoms is relatively steep off the Washington coast compared to the one along the Oregon coast. They decided on two areas, one off the Washington coast just south of where Cruise 9 ended a year before, west of Destruction Island, and the other area between Cape Lookout and Cape Foulweather just north of Newport, Oregon, which is the widest area between 100 and 500 fathoms. The continental slope is gentle in this area and appeared suitable for trawling, while off the Washington coast it is much steeper, although still possible to trawl (Drawing 5).

The Washington area was found to be less trawlable than anticipated. Eight hauls were made at depths of 100 to 310 fathoms, and only three were trawlable. Haul 45 made at a depth of 100 fathoms had a catch of 1,000 lb. of POP. The other two, haul 41 and 42, were made in much deeper water of 304 to 308 fathoms. The total catches were 400 and 500 lb., but there was no POP in either of them. Both were dominated by two species, sablefish and the idiots.

The Oregon coast was much more trawlable than the Washington area. A total of thirty-nine hauls were made, only seven unsuccessful. The remaining thirty-two hauls were spread out over a depth range of 100 to 400 fathoms, grouped into three different depths: 100-199, 200-299, and 300-399 fathoms. The hauls were generally one hour in duration, although a few were two hours. Their catches were cut in half so that they would be comparable. It must have been rewarding for Lee to observe the results of the catches made in this trawlable area, covering depths out to 400 fathoms, whereas during Cruise 9, they only went out to 225 fathoms.

The thirty-two successful hauls were divided into three depth groups: eleven in the 100-fathom group, eighteen in the 200 group, and three in the 300 group. In the 100-fathom group, there were eleven successful hauls. The best catch of 2,000 lb. of POP were in hauls 22 and 23 in 120 fathoms and haul 30 in 139 fathoms. The catch of POP in the other six hauls ranged from 20 to 800 lb. in water depths of 100 to 170 fathoms. Two hauls had no POP—38 at 120 fathoms and 37 at 199 fathoms. In the 200-fathom

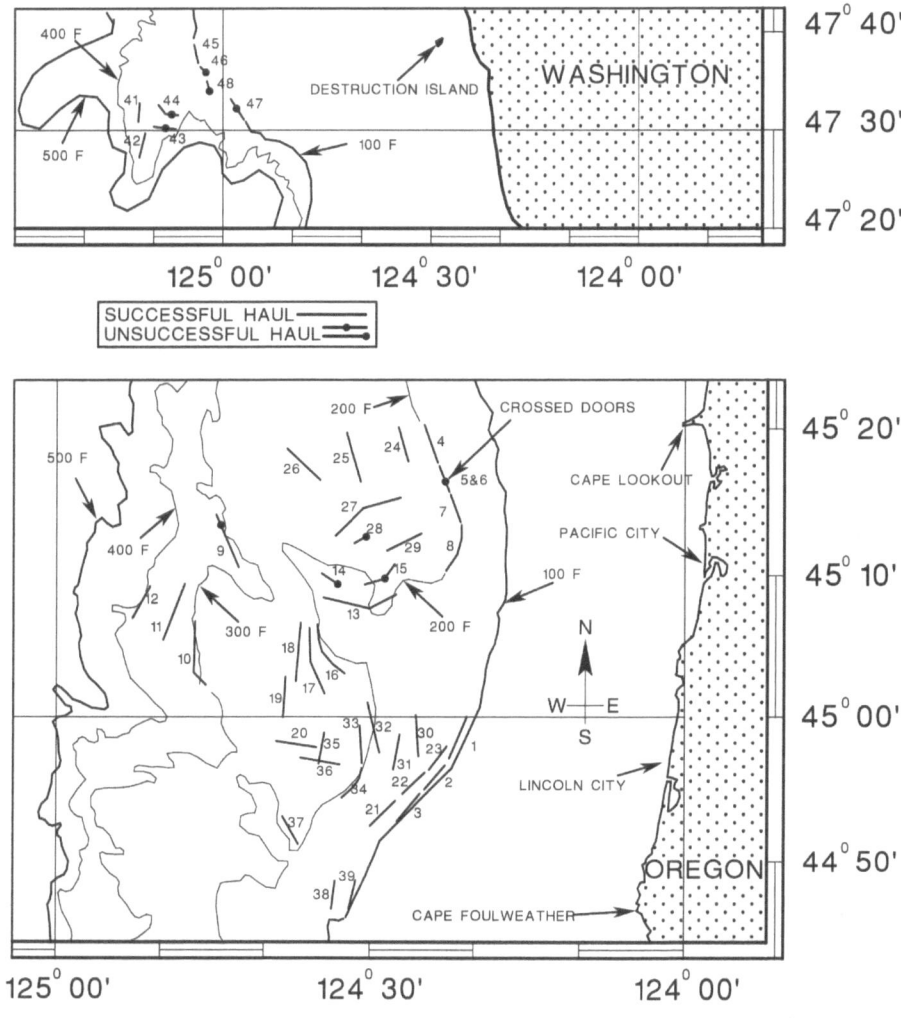

Drawing 5 R/V *John N. Cobb*—Cruise 13 Hauls.

group, there were eighteen successful hauls. POP catches were found in fifteen. Two hauls—16 in 200 fathoms and 19 in 230 fathoms—had the best catch of 1,000 lb. of POP each. The remaining thirteen had catches that varied between 20 and 500 lb. of POP in depths of 200 to 238 fathoms. Three hauls—20, 35, and 36 made at 240 fathoms—had zero POP in their catches. Haul 26 was the deepest made in the 200-fathom group, at 255 fathoms, but the catch of POP was less than 20 lb.

The three successful hauls made in the 300-fathom group indicated they were in the lower end of the POP depth range. Only haul 10, made at 302 fathoms, yielded 50 lb. of POP, whereas the other two, hauls 11 and 12, made in water of 355 and 397 fathoms, were void of POP. The depth that POP was found increased to 302 fathoms, with the best catches occurring between 120 and 230 fathoms. There were five red rockfish species taken during the cruise—*S. crameri, S. diploproa, S. saxicola, S. introniger,* and *S. rosaceus. S. saxicola* was added to the list from Cruise 9. The idiots, *Sebastolobus alascanus*, were also included in the red rockfish grouping and was found out to the deepest haul made at 400 fathoms.

Change in Job (1954–1959)

At the end of WWII, Japan and the US made an agreement that the Japanese could catch salmon on the west side of the International Date Line but not on the eastern side. The US felt that the Bristol Bay run never crossed the Date Line. The US fishermen fishing the returning adult run to Bristol Bay complained that the Japanese high seas fleet were catching sockeye from Bristol Bay since they were finding gill net marks on the returning salmon in Bristol Bay. If true, they were fishing on the eastern side of the Date Line. In 1953, the *Cobb*'s Cruise 16 was assigned to explore for salmon, a study for the Fish and Wildlife Service and the International Convention for the High Seas Fisheries of the North Pacific Ocean. The work was carried out in the offshore waters adjacent to the Aleutian Islands, Alaska. The results were published in May 1954.[15] It became a national crisis. The International North Pacific Fisheries Commission requested the Exploratory Fishing Base conduct a salmon survey of the North Pacific Ocean from Attu to Vancouver Island, Canada. The *Cobb* departed on Cruise 23 along with two charter vessels, *Mitkof* and *Paragon,* to see where the five species of salmon went in the Pacific after they left the streams of North America. The results were published in July 1957.[16] I think that Lee Alverson wasn't happy the way the vessel was being used.

Another problem for the exploratory group was the growing tendency to minimize the scientific components of the work and focus instead on the commercial value of the surveys. Alverson was more interested in bottom-fish research and especially not salmon, so he resigned from the exploratory group and applied to the State of Washington groundfish group and was accepted in mid-summer of 1954.[17]

New Director (1959)

Joseph Ellson was the first director of the Exploratory Fishing and Gear Research Base when it was created in 1949, and he served in the job until 1950. Then Donald Powell was the director from 1950 to 1951 and again from 1955 to 1958. He had the opportunity to move up to a new job in Washington, DC, with the exploratory program. He offered his old job to Lee Alverson. It was an unexpected offer, and Lee accepted it and became the new director in 1958. He had an understanding of the Washington trawl fisheries and had gained experience while he worked for the exploratory base. The concept of exploratory fishing was growing nationally.[18] While he was with the state, between 1954 and 1958, he had gotten to know the commercial trawl fishermen by developing a voluntary fishing log for each of the otter-trawl vessels for the State of Washington. It was a volunteer program he put into effect by visiting each vessel and talking with the skipper.[19] As far as I know, the log is still in use today.

Lee Alverson introduced me to Don Powell when I was hired in 1960. I was told that he had a lot to do with the early development of the base. Because of his early death on October 10, 1961, I never got to know him, and there is little information available about him. He went to Highline High School in Seattle, enrolled in Washington State College, then into WWII, was discharged, then finished up his education at the UW College of Fisheries in 1948 and started work with the US Fish and Wildlife Service in 1949.[20] When Lee became director in the summer of 1958, he was confronted with the AEC exploration of the Chuckchi Sea. Cruise 43 was scheduled to sail on July 22, 1959. While the vessel was in the shipyard preparing for the cruise, a crewman's hand went through the side of the ship while he was preparing it for fresh paint. There was extensive dry rot. I remember that summer since it was the talk at the College of Fisheries. I went down to the shipyard and saw the *Cobb* out of the water with her planking off the side so you could look through the ship. Three ribs had to be replaced. You can imagine the concern of getting the *Cobb* ready for sea in time to travel through the Gulf of Alaska and the Bering Strait to the Chuckchi Sea.

Chapter 2: Endnotes

1. Dayton L. Alverson, *Race to the Sea* (New York: iUniverse, Inc., 2008), pp. 223–228. Harold A.

2. W. A. Clemens and G. V. Wilby, "Fishes of the Pacific Coast of Canada," *Fisheries Research Board of Canada Bulletin* no. 68 (1961).

3. Alverson, *Race to the Sea*, pp. 241. New job.

4. Dayton L. Alverson and Arthur D. Welander, "Notes on the Scorpaenid Fishes of Washington and Adjacent Areas, with a Key for Their Identification," *Copeia* 1952, no. 3 (September 1952), pp. 136–143. MS #10.

5. Milton S. Love, Mary Yoklavich, and Lyman Thorsteinson, *The Rockfishes of the Northeast Pacific* (Los Angeles: University of California Press Berkley, 2002), pp. 9–11.

6. Love, *The Rockfishes of the Northeast Pacific*, pp. 6–7.

7. Charles R. Hitz, "A Trip to Nanaimo and a Last Visit with Jergen Westrheim," *Carmel Finley* (blog), December 11, 2012, https://carmelfinley.wordpress.com/2012/12/11/a-trip-to-nanaimo-and-a-last-visit-with-jergen-westrheim/. Bob's posting #6.

8. Frank E. Firth, *The Encyclopedia of Marine Resources* (New York: Van Nostrand Reinhold Company, 1969), pp. 579-583. Redfish fishery by G. M. Clarke.

9. Love, *The Rockfishes of the Northeast Pacific*, pp. 6–7. Common names.

10. Love, *The Rockfishes of the Northeast Pacific*, pp. 6–7. East Coast species

11. Alverson, *Race to the Sea*, pp. 242–243. Schedule.

12. Dayton L. Alverson, "Deep-Water Trawling Survey off the Coast of Washington (August 27–October 19, 1951)," *Commercial Fisheries Review* 13, no. 11 (November 1951). MS #11. See also Sep. No. 292.

13. Love, *The Rockfishes of the Northeast Pacific*, pp. 122. S. introniger = S. aleutianus.

14. Dayton L. Alverson, "Deep-Water Trawling Survey off the Oregan and Washington Coasts (August 25–October 3, 1952)," *Commercial Fisheries Review* 15, no. 10 (October 1953). MS #18. See also Sep. No. 35.

15. Edward A. Schaefers and Francis M. Kukuhara, "Offshore Salmon Explorations Adjacent to the Aleutian Islands, June-July 1953," *Commercial Fisheries Review* 16, no. 5 (May 1954). MS #26. See also Sep. No. 371.

16. Donald E. Powell and Alvin E. Peterson, "Experimental Fishing to Determine Distribution of Salmon in the North Pacific Ocean," *Special Scientific Report—Fisheries* no. 205 (July 1957). MS #31.

17. Alverson, *Race to the Sea*, pp. 270-272. Job change.

18. Alverson, *Race to the Sea*, pp. 320-321. Director.

19. Dayton L. Alverson, "Study of Annual and Seasonal Bathymetric Catch Patterns for Commercially Important Groundfishes of the Pacific Northwest Coast of North America," *Pacific Marine Fisheries Commission Bulletin* 4 (1960). MS #51.

20. "Donald E. Powell," *Seattle Times*: Obituary (October 10, 1961).

CHAPTER THREE

College Years 1951–1960

I graduated from Bellingham High School in Washington in 1951 and was drafted into the US Army on October 6, 1954, after attending Washington State College (WSC) in Pullman, Washington, for three years (WSC became a university in 1959). I was discharged on August 29, 1956, in time to register for my fourth year at WSC. After completing the school year, I married Maureen in June 1957, and we decided to move to Seattle. The opportunities for employment in Seattle were far greater than in Eastern Washington. Seattle was closer to my hometown, Bellingham, further north up the coastline of the Salish Sea. If I were to be honest, I liked fishing and boats too much to be too far away from the water.

I was so close to getting my zoology degree, so we decided I should register at UW in Seattle and transfer the credits I needed to graduate from Pullman. So, we rented an apartment near the UW campus, Maureen got a job in Seattle, and I got a summer job with the Seattle Parks Department and had time to look for a future job as a zoologist. No one seemed interested in hiring a zoologist, until I visited the Montlake Lab near the Seattle Yacht Club. There, I was told I could start as a biologist for them as long as I took a few basic fisheries classes at the University of Washington next door. Finally! I thought that this sounded great. They were interested in a zoologist. The classes I had taken at WSC in Pullman weren't wasted, and I was already registered at UW to get the few credits I needed to finish up my bachelor of science in zoology. I had always been interested in fish and had spent a lot of time on the water growing up on Bellingham Bay and in the San Juan Islands. The path to this moment had been winding, and I still wonder if it was a question of chance or divine guidance that led me to this moment. Soon after I began my classes, UW had a guest speaker, Dayton L. Alverson. He had just returned from a United Nations Food and Agriculture Organization conference on world fishing methods held in Hamburg, Germany, in October 1957. He spoke about the distant water trawling fleets of the world and the major changes that were beginning to occur with the introduction of stern trawlers, which would make the current side trawlers obsolete.

He had the opportunity to go aboard a new Russian factory stern trawler,[1] one of forty-two 277-foot Pushkin class stern trawlers built in West Germany between 1954 and 1956.[2] It was based on the new design of the first factory stern trawler, *Fairtry*, launched in Scotland in 1954 by a Scottish whaling company. The unique stern ramp used in the whaling industry was used in the design of the factory trawler. The catch could be dragged up the stern then dumped to the factory below the deck for processing.[3] These new stern trawlers were large vessels, and their different methods of handling nets and catches fascinated me (Fig. 14).

Fig. 14 Soviet Pushkin Class Stern Trawler. Image #3055.

One of the required classes was fish taxonomy. I found it extremely interesting, especially when we got into the large group of fishes referred to as rockfish. There are sixty species of *Sebastes* living along the northeastern Pacific coast of the US today, while there are only four species in the Atlantic Ocean.

These were the same rockfish I'd caught in the San Juan Islands when fishing for salmon during my grade school and high school years. Seen near the surface, these fish were a dark color. We called them sea bass or rock cod. When we fished deeper, they were bright red and were called red snappers. At the university, I learned that red was the first color removed from the spectrum of sunlight when it penetrated water, so a red fish is like a black fish in deeper water, having an ideal method of camouflage. Their method of reproduction is classified as ovoviviparous, or giving birth to live young. I'd had the idea that all fish released their eggs outside the body and were fertilized externally. Although I had filleted many rockfish, I'd never opened the body cavity to check the eggs until UW.

The classes and atmosphere at UW were so engaging that instead of trying to get employment, I decided to apply to graduate school in marine biology. I was accepted and began studying for a master's degree. Financially, Maureen and I were comfortable, and employment wasn't on my mind. I had the GI Bill, and she was employed in Seattle. We had an apartment within walking distance of the UW College of Fisheries, and, finally, I was supported with a Washington State research grant for my graduate work on two different rockfish in Puget Sound, the brown rockfish (*S. auriculatus*) and the copper rockfish (*S. caurinus*).

Fig. 15 R/V *Commando*—University of Washington Research Vessel. Image #4037.

UW had a research vessel, the *Commando* (Fig. 15), which is a Pacific coast combination vessel, that I would use to collect my samples. It's unbelievable, thinking back, that I had the use of a commercial trawler, one that had been used in the halibut longline fishery for eleven years and was sold to the UW College of Fisheries in 1955, replacing their R/V *Oncorhynchus*. I would capture the samples I needed to carry out my project by using the *Commando* to tow an otter trawler in Port Orchard, Washington.

The *Commando* had an overall length of 67 feet and was built for the commercial fishing industry in Seattle in 1944. It was a typical Pacific coast combination vessel, rigged for trawling, and was my first introduction to a commercial fishing boat. Trawler, dragger, and otter trawler are all terms that define a vessel that tows a trawl net along the bottom. Trips on the *Commando* were a full day. During 1959 and 1960, we made a total of twenty-two—two or three during the winter, once a week during the spring, and then only once in the summer, with a couple more in the fall.[4] There were three of us on the vessel: Tom Oswald Jr., the skipper, Olaf Rockness, a man of many talents—the engineer, deckhand, and part-time cook—and then there was me, a greenhorn deckhand. It was a wonderful experience, and I still remember it clearly.

We departed from the UW College of Fisheries' dock located just west of the Montlake Cut across from the Seattle Yacht Club, casting off the lines in the early morning and heading out, blowing the whistle to open the University Bridge. After that, we followed the ship canal to the Hiram M. Chittenden Locks, where we waited to enter. Once inside, we waited again for the after-lock door to close behind us. Then we waited once more (and became a tourist attraction) as water was pumped out of the locks, lowering the *Commando* to the level of Puget Sound on the other side. Once the water reached that level, the outer doors would open and we would move out into Shilshole Bay and salt water, heading across Puget Sound, bound for Port Orchard. While we crossed, Olaf would go into the galley to fix us breakfast on the galley stove. It's been over fifty years since I was served breakfast in the *Commando's* galley, and I still remember Olaf's biscuits were with bacon and eggs. After breakfast, we would continue heading for Agate Passage. Once there, we would go under the bridge on the northwest side of Bainbridge Island. Agate Passage opened into a body of water called Port Orchard, which had a smooth bottom where past trawls had yielded rockfish. We would set the trawl and tow it for an hour (Drawing 6).

Since there were only three of us onboard, I had to run one of the winches when we set and retrieved the trawl. There was a brake handle on each winch, consisting of a wheel that released the brake when turned. When letting out the gear, Olaf kept saying, "Make sure the brake is off and it's not dragging." So I kept unscrewing it to make sure it wasn't dragging until one day, I unscrewed it completely, and it sprang out of the socket. I said, "Is this OK?" He said a lot of bad words. How he refitted the screw

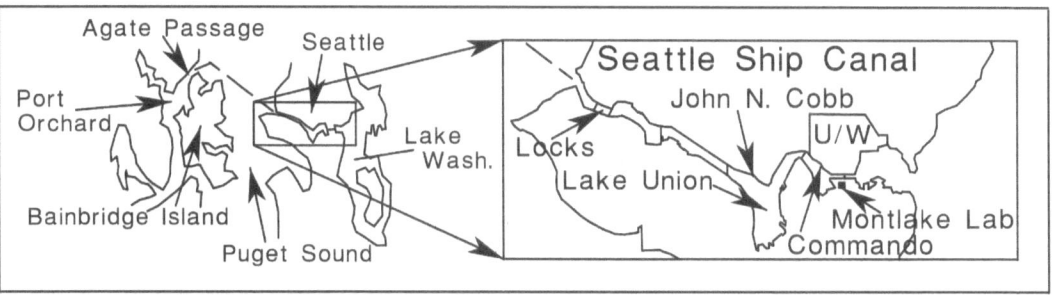

Drawing 6 Map of Seattle Ship Canal and Port Orchard.

into the socket is still a mystery to me, but he did, and I never unscrewed it completely again.

Retrieving the trawl was an interesting procedure. The cable came in and had to be laid flat in layers, so that all the cable could be contained on the drum. Each winch operator had an iron bar he used to guide the cable evenly onto the drum. I also had to take in the single line from the boom that was used to lift the catch aboard. Olaf told me never to let go of it until it was fastened or passed to another crewman, because when the vessel was outside and rolling, a lost hook was dangerous. A few years later on the *Cobb,* I had an opportunity to see one that got loose. Olaf was correct; it was terrifying.

Once the net was brought back aboard and the catch sorted, the rockfish were set aside. On the way back to UW, I measured each for total length and opened the body cavity to determine its sex and what stage the females were in. Rockfish are ovoviviparous, producing eggs that develop within the ovary and are released into the sea as larvae. I found that the ovaries changed color as the embryos grew—from yellow to orange as the yolk is absorbed, gray to black as the pigment is laid down on the embryo, and black with small golden spots as eyes developed and the yolk sac disappeared. As the larvae hatch and the yolk sacs disappear, the young larvae are released to the sea to fend for themselves at the rate of tens of thousands per female each year.[5]

During my graduate work, my advisor arranged for me to meet Dayton L. Alverson, who was located in the same building I had visited two years before when I was encouraged to take a few fisheries classes. He was the new director of the branch of the US Fish and Wildlife Service called Exploratory Fishing, and I was to talk to him regarding his knowledge of rockfish, the subject of my research. I went to the second floor of the building, was ushered into his office, introduced myself, and told him why I was there and that I wanted to know more about the rockfish.

Alverson told me about how the POP (*S. alutus*) had become dominant in commercial fisheries, and he believed there was a vast untapped resource of them in the deep waters off our coast. We discussed other aspects of that group and the timing of their spawning. He also talked about the *John N. Cobb* finding an unknown seamount off the coast of Washington, a mountain that rose up near the ocean's surface, called the Cobb Seamount. Longlines had been set across it and had come up with several large rockfish, called red snapper, *S. ruberrimus*.[6] Years later, a book was published describing Project Sea-Use, the exploration of the top of the Cobb Seamount in 1966 with the use of scuba divers.[7]

The information Alverson supplied was both fascinating and useful. I took copious notes in a new notebook. As we finished, I got up to leave, and he came out from behind his desk, reached out, and snatched my notebook! He then retreated to his desk and began reading my notes, proceeding to announce that I had "misspelled this fish and that fish" and on and on and on. To say I was embarrassed would be an understatement. I had always been a very poor speller, and to have it so clearly and publicly pointed out was my worst nightmare. I finally left with a red face and corrected notebook, feeling that if ever there was a job with Exploratory Fishing, I would most likely be the last to be considered.

In 1960, I and two other graduate students had the opportunity to accompany the *Commando* for a week's trip to the open ocean to collect true cod off the coast of Washington. It was my first opportunity to encounter the open waters of the Pacific Ocean. Before I left on that trip, I wondered if I would become seasick. There was no class about seasickness given at the College of Fisheries, but there was considerable talk.

Seasickness is strange. Some people get it, and others don't. I thought I wouldn't because of the storm I went through on the Atlantic Ocean on my way back from France in 1956. I had completed my active duty with the US Army and was headed for discharge in Seattle on a 623-foot troop transport, the USS *General H. W. Butner*, when she was caught in a hurricane. The seas were monstrous. The vessel was hove to, and I, along with the rest of the troops, was locked below decks. I remember lying in one of the bunks near the after part of the ship, and as abnormal swells passed under us and the bow hung in the air, the whole ship would vibrate from bow to stern. We hoped the welds would hold so she wouldn't split in half. I had guard duty during part of that time, and the sergeant put me in the middle of a stairway in the bow that let out onto the foredeck, where it was my job to keep anyone from going up on deck. At the bottom of the stairway was a head full of vomit and some soldiers throwing up. As I sat on the stairs, we would rise 20 to 30 feet as a swell approached and then drop the same distance as it passed underneath us, before coming back up to repeat, again and again. Fortunately, nobody came up the stairs. I was proud I never got sick, though I did feel woozy. After that experience, I thought I was immune.

We left the college on Monday morning, May 2, 1960, for the weeklong trip. By dinnertime, we were off Dungeness Spit and heading out the Straits of Juan de Fuca. A strong westerly wind was blowing directly down the strait with an ebbing, outgoing tide against the wind, making the waves even steeper as the vessel bucked into them. Since I was in the pilothouse when dinner was called, I went back to the galley and sat down to eat, but after a bite or two of spaghetti, I had to rush out the galley door to the deck—I was seasick. The motion of the *Commando* in that chop really got to me.

On deck, I threw up and leaned against the mast. Looking aft at the horizon from the center of the vessel made me feel better, so I went back to the pilothouse and remained there until I started feeling sick again. Back I went to the mast, hoping that would help as it had before. As I looked around, I saw that the deck was covered with vomited spaghetti. How could I have thrown up that much when I'd eaten so little? Dave, one of the other students, later said that he had eaten a full dinner and then some and lost it all on deck. The third student gave us a bad time about seasickness since he was feeling great and felt that the sea was his calling.

We put into Neah Bay that night and proceeded offshore the next morning. The day was nice, with a slight breeze and a swell from the northwest, and my seasickness was somewhat under control. Olaf gave me a broom and told me to sweep the deck. The open air helped keep my mind off thinking about being sick. Dave was also sweeping the deck, but the third student was seasick off and on throughout the remaining days off the coast. As soon as we returned from the trip, he changed his major to teaching and did a wonderful job as a science teacher on dry land once he graduated. We made four hauls the first day out, each about two hours long and in depths of 57 to 61 fathoms along the so-called bread line, as the fishermen call it, just south of Cape Flattery. After the fourth haul was up at 8:00 p.m., we

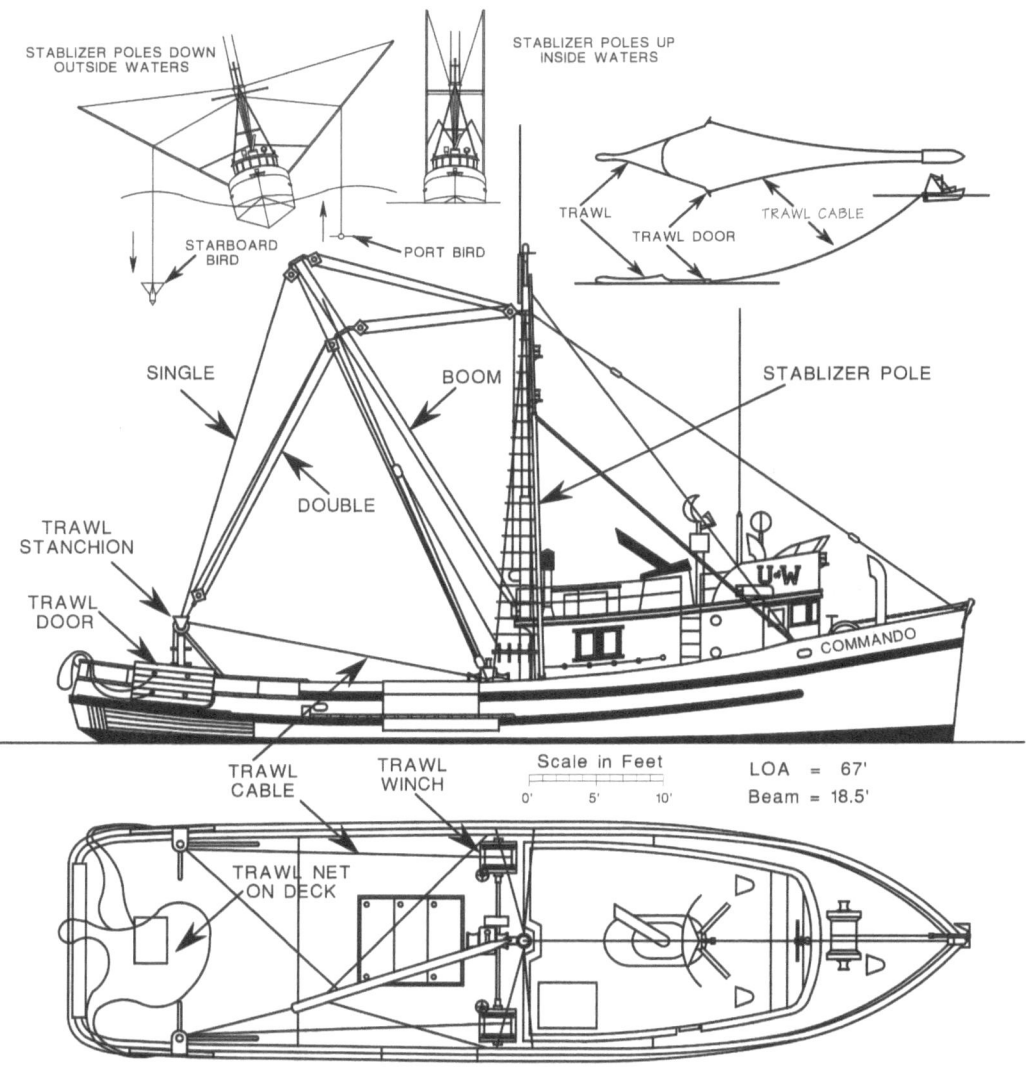

Drawing 7 R/V *Commando*—Configuration (1960).

secured everything and shut down for the night somewhere near Destruction Island. We had taken twenty-five true cod in the first three sets and seventy in the last one. There were a few rockfish in the catch, and I opened each to see the condition of their ovaries. Then everyone went to bed. I was in the upper bunk on the starboard side of the house.

I lay in the bunk, hearing a swishing sound that mystified me. The *Commando* was very quiet with everyone in their bunks. The main engine and the auxiliary were shut down, the little power needed supplied by a bank of batteries. The noise came from outside, first from one side and then from the other, as the vessel was moved by the gentle everlasting swell that was present on the open coast. It finally dawned on me it was the sound of the stabilizer cables as they moved through the water. I had watched them being set when we left Neah Bay for the open ocean. The poles stored in the upright position while in enclosed waters are let out to about forty-five degrees on each side. From each pole, a cable is attached with a bird at the end, which is submerged below the waterline. The bird is a weighted cylinder with a triangular fin on each side, like the wings of a delta jet, and a vertical tail. It is designed to help reduce the roll of the vessel when at sea.

I assumed we were drifting and should have gone to sleep because the noise was so soothing. I had seen lights on the beach before we went to bed, which meant we were close to shore, and I was concerned we could drift ashore. Tom and Olaf slept soundly. Nobody was on watch. Morning came, and everyone got up and went into the galley, where we waited for the coffee to perk. Looking out the door, it seemed as though we hadn't drifted during the night, and puzzled, I mentioned it. No one said a word. The coffee continued perking, and as they were drinking it, I looked out again and said, "That spot on shore is in the same place as it was last night, so why didn't we drift? We should have." No one said a word. When Tom and Olaf had finished their coffee, Tom told Olaf, "Start up the engine and then haul the anchor," and that's when I finally realized we had anchored all night in the open ocean. I had thought that no one in their right mind would ever anchor in the open ocean. If I had known it was common practice, I would have slept a lot better. Talk about a greenhorn!

After the anchor was pulled that morning, we proceeded offshore and made our first set at 4:50 a.m. in about 58 to 60 fathoms for a two-hour tow south near Destruction Island. We took forty true cod in that haul and then made a second haul in about the same location at 7:22 a.m., which lasted another two hours and took another forty true cod. Again, there were a few rockfish in the catch, which I measured and sexed, along with a few flatfish and small halibut. There was a surprise in the catch—a white sturgeon. We all took part of it home as home pack. It was delicious. On the way back, we made our last haul in Puget Sound at Port Orchard at 4:35 a.m. the next morning, May 5, 1960, the last haul made in my study. It is recorded in the *Commando* fish log as haul 6017-G, the seventh tow for the trip. I had survived my first coastal trip and realized that seasickness was just part of the job.

My thesis was about completed. One item holding me back from my master's degree was the requirement to pass a UW French exam. When I was overseas in France, I had enrolled in a French class, which was used for my undergraduate language requirement. But in graduate school, the exam was to convert a story in French into English. I tried at least six times without passing. I was frustrated.

Lee Alverson offered me a job in the Exploratory Fishing group toward the end of May 1960, which I believed was due to my interest in rockfish. I couldn't believe he offered me the job. It was one I wanted!

Chapter 3: Endnotes

1. Dayton L. Alverson, *Race to the Sea* (New York: iUniverse, Inc., 2008), pp. 314. Pushkin.

2. Charles R. Hitz, "Catalogue of the Soviet Fishing Fleet," *National Fisherman Yearbook Issue, 1968* 48, no. 13 (March 1968), pp. 9–19, 21–24. MS #172.

3. James Strong and Keith R. Criddle, *Fishing for Pollock in a Sea of Change: A Historical Analysis of the Bering Sea Pollock Fishery* (Fairbanks, AK: University of Alaska Fairbanks, 2013), pp. 4.

4. Allan C. DeLacy, Charles R. Hitz, and Robert L. Dryfoos, "Maturation, Gestation, and Birth of Rockfish (*Sebastodes*) from Washington and Adjacent Waters," Washington Department of Fisheries, Fisheries Research Papers 2(3) (1964), pp. 51–67. MS #62.

5. Charles R. Hitz, "*R/V Commando*—College of Fisheries—Puget Sound," *Carmel Finley* (blog), March 17, 2015, https://carmelfinley.wordpress.com/2015/03/18/rv-commando-college-of-fisheries-puget-sound/. Bob's posting #30.

6. Alverson, *Race to the Sea*, pp. 252-256. Cobb Seamount.

7. Bill Brubaker, *Seamount: Discovery and Exploration of Cobb Seamount* (N.p.: CreateSpace Independent Publishing Platform, 2016).

CHAPTER FOUR

Cruise 46, Spit 1960

The new job, which I started on July 5, 1960, was with the Exploratory Fishing and Gear Research Base located at the Montlake Lab and was in the original brick building. I had found an article in their Research Library entitled "Exploratory Fishing" by H. E. Crowther in the *Fishing Gazette*, 1949 Annual Review. It's hard to describe the excitement I felt as I read it.

The article explains the concept of exploratory fishing: "It is logical, therefore, that off-shore exploratory fishing operations should be carried on by the Government, but it was not until recently that the Federal Government began developing an exploratory fishing program....During the years in which exploratory fishing has been neglected, this country has conducted two types of scientific investigation in the fisheries. One is biology, the other technology....Between the biological studies of fish on known fishing grounds and the technological research on handling fish after being caught, there is a definite gap, the science of: (1) locating new fishing areas, and (2) determining the most effective means of catching fish wherever located." The article also explains the makeup of crew members aboard exploratory vessels. "The crew members aboard an exploratory fishing vessel are not unusual individually, for each is an experienced commercial fisherman, but as a crew they are unusual, for their total experience must include knowledge on every important species in the area being fished....In addition to the captain and crew, the exploratory vessel carries specialized scientific personnel. These specialists are known as Fishery Engineers. Through education, training, and experience they must be qualified to plan and direct various phases of exploratory fishing programs, design and direct operation of fishing gear, evaluate and record results of fishing tests and identify the species of fish caught."[1] This explains why I was hired as a Methods and Equipment Specialist, the first step to becoming a Fishery Engineer. Two or three years later, our titles were changed to Fisheries Biologist.

By 1960, the Montlake Laboratory housed three groups: Biological Lab, Technologial Lab, and Exploratory Fishing and Gear Rearsearch Base. They were part of the Pacific Region of the Bureau of Commercial Fisheries.[2] I applied for a job with the third group, Exporatory Fishing. During the rest of May and June, I filled out the paperwork and supplied the documents to get hired by the federal government. I anxiously waited to hear from them and was finally told to report to work in July 1960. I was hired in the job I wanted. It was a new concept. I was aware of how the *John N. Cobb* was being used to discover new fishing grounds in order to get fishermen to harvest other resources than the salmon and halibut, which were the mainstay fisheries of the West Coast in the latter part of the 1950s. I was interested in exploring the bottom of the ocean and learning more about POP.

In 1958, Dayton L. Alverson became the director. He still had an intense interest in this fascinating group of Pacific rockfish, which is why I believe he picked me up in this whirlwind expansion of exploratory fishing during early 1960. When Lee returned to the exploratory group after working for the State of Washington, he continued where he had left off six years before with the added experience he gained while working with different commercial fishermen. Based on those friendships, he learned where the *Cobb* should explore for bottomfish. They told him about areas like the Spit, where the bottom was too rough to fish on the continental shelf, and mentioned that once in a while, when they tried some of those areas with an old net, if they got the gear back at all, there were commercial fish in it. He should also explore along the top of the slope deeper than 100 fathoms to determine what species lived out there. Lee used this information in planning future bottomfish explorations of the *Cobb*.

Lee continued visiting the skippers of the commercial trawl fleet after he became director, and after I was hired, I accompanied him on one of the visits. One day, we were coming back from a meeting, driving along the Seattle waterfront. He saw a trawler alongside the Main Fishing company's dock, which was a landmark in those days, and said we needed to stop, so we parked and walked along the wharf until we stood above her. The crew was cleaning up after unloading fish. Lee called down, asking if the skipper was aboard. When they said he was, Lee asked if we could come aboard. They said yes. We went down a ladder that hung from the top of the pier and stepped onto the fishing deck of the trawler just as the skipper came to the galley door. He saw us and swore as only fishermen could, then asked, "What do the bug pickers want this time? Probably coffee, so come on in." I was quiet, wondering what we could learn from such a negative reception. We sat at the galley table as the coffee was poured, and he and Lee began to talk about where he had been fishing, what was going on in the fishing business, how the fishing log was working, and what was happening in Washington, DC. It became a very friendly meeting. When we left, the skipper said, "Stop by anytime. It was great to see you again."

Alverson became the third director of the base in the summer of 1958, sometime around the first of May when the *Cobb* was involved with the exploration of shrimp for the rest of the year. The next cruise, 43, was an important one. They were to gain information for the proposed AEC experimental atomic blast to create a harbor in the Chuckchi Sea. I remember that period well since some of the College of Fisheries students and professors had temporary jobs on the trip. The conversations around school were about the repairs of the dry rot that was fortunately found during her annual shipyard repairs, where the funds were coming from to repair her, and whether she would be ready to sail on time. I can only imagine the concern that the new director had during that time.

The *Cobb* sailed on July 22, 1959, with the director onboard. Before crossing the Arctic Circle, they ran into a Russian distant water fleet, fishing with otter trawls in the eastern part of the Bering Sea off Kuskokwin Bay, AK, just north of Bristol Bay, on August 5, 1959.[3] Then the *Cobb* proceeded to the Chuckchi Sea and conducted the survey, returning to Seattle on September 11, 1959, completing the Chuckchi trip.[4] Two additional *Cobb* trips were made before Lee could get his third rockfish survey scheduled. Finally, Cruise 46 was scheduled for a bottomfish exploration of the Spit area off the Washington and British Columbia coasts. It's ironic that Cruise 46 started on Monday, May 2, 1960, the same day I sailed on my first ocean cruise off the Washington coast on the UW *Commando*. We never saw the *Cobb*. I'm sure she passed us when we were in Neah Bay for the night. I had no idea I would be involved with Cruise 46. After I was hired and completed my first *Cobb* cruise, I was involved in writing the end reports for both Cruise 46 and 47, which were combined into one.[5]

I have not included the Otter Trawl Fishing Log for Cruise 46 & 47 that came with some of the reprints of the report. Since 1951, the POP market was growing, and the exploratory base took the species out of the Red rockfish group and treated it separately as POP since the fishermen and the fish houses were interested in the catches of that induvial species. The Otter Trawl Fishing Log attached to the main report, Separate No. 620, which lists all the net hauls made during Cruise 46 and 47. They were separated out from the station number and renumbered chronologically from haul one of Cruise 46 to the last, haul 56, made on Cruise 47, excluding all the other station numbers, such as sounding transects and chain drags. The original field data sheets for both the cruises are stored at the Seattle National Archives.[6] The only way to get the individual species that are grouped as red or black rockfish is to go back to the original station number where the species of rockfish are recoded, which I did recently. After Lee was hired as the director in 1958, each species of rockfish taken in each haul had to be identified to species. We were instructed to record them in the original field data sheets by station number, even though they are not shown in the published log. I recorded each species of rockfish by their scientific name and grouped them into two groups, black rockfish and red rockfish and included POP in the red group. They are now listed by the net haul number.

Cruise 46 — The Spit

Lee finally had his third rockfish cruise. The Spit area was covered between May 7 and June 22, 1960, before I came to work on July 5, 1960. The objective of the trip was different from the first two cruises, 9 and 13, with which Lee had surveyed waters that were 100 fathoms (600 feet) and greater in 1951 and 1952. Cruise 46 was in waters of 50 to 100 fathoms on the shelf, where fishermen concentrated their trawling efforts. They knew the dangers for damaging trawl gear on the Spit. If any successful tows were found, the results would benefit the industry and give us a better understanding of what species of rockfish were found on the shelf. If the catch of an individual species was one thousand or more pounds, it was considered large.

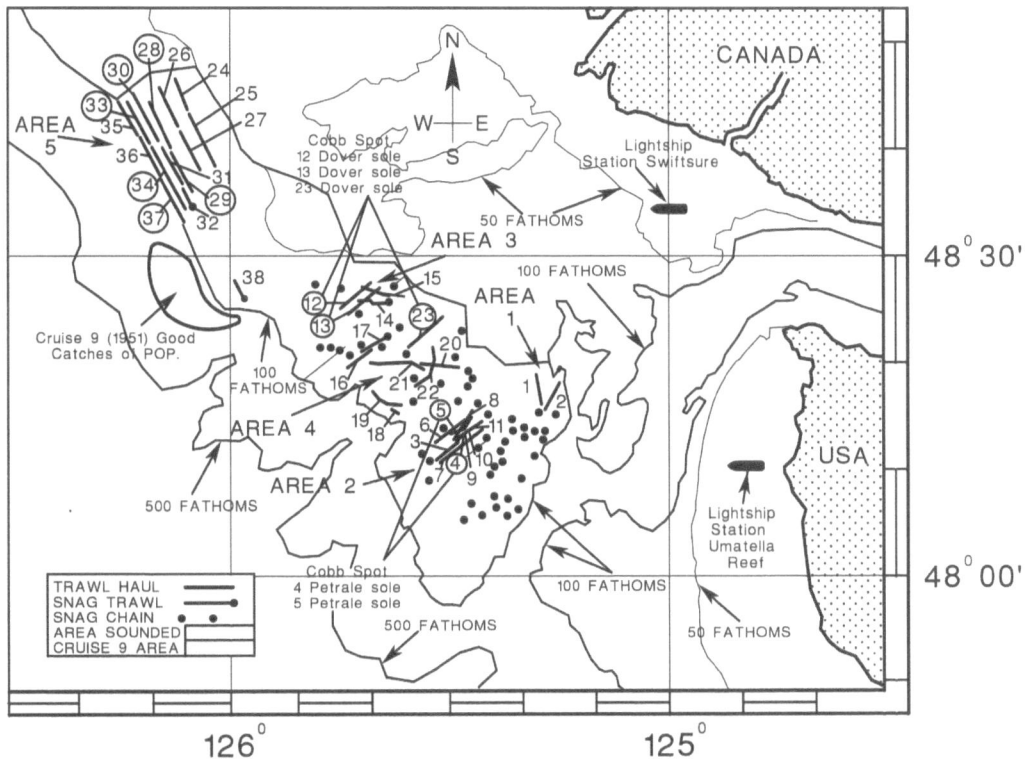

Drawing 8 R/V John N. Cobb—Cruise 46 Hauls.

This was the first cruise where the new sounder was used that divided the echo of the fish from the bottom of the ocean. It also showed whether the bottom was soft or hard. When the bottom was found to be flat and smooth, a chain was used, replacing the net between the trawl doors. If the chain did not hang up, the net was replaced and the net haul made. With time, a correlation was found that soft bottom was trawlable and hard bottom was untrawlable. This saved time and expense in repairing damaged nets, which was not done in Cruises 9 and 13. The sounding transects, the chain hauls, and the net hauls were all assigned a station number. The net hauls were made after the chain haul was successful for at least an hour, and they were renumbered from 1 to 38. A total of thirty-eight possible net hauls were found on the Spit in five areas that were determined trawlable by the use of the chain. They were made in less than 100 fathoms of water, all on the continental shelf, after sounding the Spit for soft bottom (Drawing 8). Four hung up—14, 17, 32, and 38—leaving thirty-four successful trawl hauls.

The published otter-trawl fishing log, which accompanied the US Department of Interior Fish and Wildlife Service Separate No. 620, lists the haul numbers chronologically and were used in the final publication. Hauls 1 to 38 were made in water depths of 62 to 90.5 fathoms. Every haul has two depths: the starting depth and the ending depth. By adding them together and dividing by two, you get an average depth. For example, haul 12 had a depth range of 72-59. Added together, that is 131. When we divide by 2, it equals 65.5 fathoms. The catch in the published log are listed by four groups: Flatfish, Round fish, Rockfish, and other. Table 1 portrays the way the data is presented in the log. Five hauls—4, 5, 12, 13, and 23—were selected because of the large catches of two valuable species of flatfish—Dover and petrale sole—and were reported by radio to the trawl fleet working nearby. A heavy border is drawn in the table around the species of a catch of 1,000 lb. or more, and the total catch in lb. is shown in the center of the square.

Before the survey was made, the fishermen Lee spoke to wanted to know about the location of any haul that yielded commercial quantiles of petrale sole and Dover sole on the Spit right away. Hauls 4, 5, 12, 13, and 23 had good catches of both species. After the *Cobb* left the area, the skipper called several trawlers nearby and gave them the location and their catches. The fleet landed over three hundred thousand pounds of petrale sole and Dover sole from these new tows. The location of these hauls are known as the *Cobb* spots and are circled in Drawing 8. Hauls 4 and 5 had large catches of petrale sole, and hauls 12, 13, and 23 had large catches of Dover sole. This was a major find for the Exploratory Fishing Base!

In the published log, the individual rockfish species are not listed, only grouped into black rockfish and the red rockfish with the exception of one species of red rockfish, POP *(S. alutus),* which is listed by itself. The only way to find the name of the species of rockfish taken in a haul is to go to the actual data sheets stored at the Seattle National Archives. In Table 1, haul 12 had 100 lb. of black rockfish made up of one species, *S. brevispinis*, and 300 lb. of red rockfish made up of three species, *S. paucispinis, S. pinniger*, and *S. rubrivinctus*.

The survey was divided into five areas. Area 1 had only two hauls, 1 and 2. Both were almost pure catches of dogfish sharks—1,500 and 8,000 lb. each—and no red rockfish or POP, with a trace of black rockfish, *S. brevispinis,* in haul 1. The hauls made in Areas 2 and 3 are all close together, especially in Area 2, in relatively small-confined areas of soft bottom. The

Table 1 Commercial Catches of Petrale Sole and Dover Sole.

Area	2	2	3	3	4
Haul No.	**4**	**5**	**12**	**13**	**23**
Date	5/13/60	5/13/60	5/28/60	5/28/60	6/3/60
Depth	80–66	79–69	72–59	72–57	72–80
Ave. Depth	73	74	65.5	64.5	76
Tow Min.	60	60	75	80	90
Total Catch Lb.	1285	2000	5550	4540	2000
Flatfish					
Dover	–	–	4000	3800	1500
English	–	–	–	TR	–
Halibut	–	TR	–	–	23
Petrale	1200	1100	–	TR	–
Rex	–	–	–	TR	–
Rock sole	–	–	–	–	–
Turbot	–	TR	1000	300	150
Roundfish					
Ling cod	25	TR	TR	50	TR
Sablefish	–	TR	TR	TR	35
True cod	–	TR	TR	TR	–
Rockfish					
Black	60	140	100	35	50
POP (S. alutus)	–	–	–	–	35
Red	–	150	300	255	40
Other					
Dogfish	–	560	75	30	100
Ratfish	–	TR	50	TR	50
Skate	–	–	–	TR	–
Hake	–	–	–	–	–

two areas are separated by Area 4, where the hauls are spread apart. Haul 23 is in Area 4 and is close to the border of Area 3. In Area 2, there were nine successful hauls made at depths of 68.5 to 79 fathoms. Two hauls, 4 and 5, had good catches of 1,200 and 1,100 lb. of petrale sole. The other large catch was haul 8, where 2600 lb. of dogfish were taken. While not a valuable fish at this time, during WWII, they were like gold due to a high demand for Vitamin A, obtained from the livers of sharks like dogfish. In Area 3, there were only four hauls made. Three were successful, made at depths of 63 to 65.5 fathoms. Hauls 12 and 13 were dominated by Dover sole in quantities of 4,000 and 3,800 lb.

Appendix 2 to 9 located at the end of the book are data that was collected for this book obtained from the original data sheets stored at the Seattle National Archives. The groups of black and red rockfish species are broken down into identified species. They are listed by haul number and compared by depth from the hauls made on the shelf to the deeper water on the slope. If the catch was 1,000 lb. or more for an individual species, the square lines surrounding the weight of the catch are highlighted. If the catch was less than 20 lb., they were recorded as a trace (TR). Appendix 2 lists the rockfish species taken in the twelve successful hauls made in Areas 2 and 3. The first two hauls made in Area 1 and haul 14, which was not successful, are not included in the appendix. The haul numbers are all on the shelf. There were no catches that were 1,000 lb. or more of rockfish in either area. Black rockfish dominated the catch with *S. brevispinis* taken in ten of the twelve hauls in catches of 25 to 480 lb. and *S. flavidus* taken in eight of the twelve hauls with catches from a trace to 220 lb. There was no POP taken in the twelve hauls, but there were five other species of red rockfish taken in them. *S. elongates* was found in two of the hauls in trace amounts, *S. paucispinis* was in nine hauls in amounts that ranged between a trace to 150 lb., *S. pinniger,* was found in eight and ranged from a trace to 100 lb., *S. rubrivinctus* was found in only three hauls that ranged from 80 to 185 lb., and *S. wilsoni* was taken only in a trace amount in one haul.

Area 4 had a relatively large area of trawlable soft bottom. The eight net hauls made there, 16 to 23, were spread out, and one of them, haul 17, was unsuccessful and not included in Appendix 3. The remaining seven hauls were made in water depths of 72.5 to 77 fathoms and were all made on the shelf. POP showed up in five of the seven hauls with a catch of 250 lb. at a depth of 74 fathoms. Four other hauls had catches of POP from a trace to 330 lb. There were two species of black rockfish. One of them, *S. flavidus,* was found in one of the seven hauls with a catch of 65 lb. The other one, *S. brevispinis,* was found in six of the seven hauls with three large catches of 1,000 lb. or more, made in hauls 16, 19, and 20, whereas the other three hauls varied from 30 to 590 lb. There were six species of red rockfish. *S. pinniger* was found in all seven hauls, 16 and 19 having a 1,000 lb. or more, three hauls with a catch of 20 lb. each, and the other two with a catch of 150 to 200 lb. *S. proriger* was only taken in one haul, 16, with a large catch of 2,500 lb. There were four other species of red rockfish taken in very small quantities: *S. elongates* was taken in one haul, 21, in a trace amount. *S. paucispinis* was taken in two hauls of 35 and 55 lb. *S. ruberrimus* was found in two hauls with trace amounts in both catches, and *S. rubrivinctus* was also taken in two hauls, one with a trace and the other 60 lb.

Area 5 in the northern part of the Spit had a soft bottom close to the 100-fathom contour, where fifteen net hauls were made. Two of them were unsuccessful, leaving thirteen successful hauls made in water depths of 62 to 90.5 fathoms all on the shelf but coming close to the continental break at 100 fathoms. Appendix 4 lists the rockfish taken in the thirteen successful hauls made in the area. POP were found in eleven of the thirteen, starting at haul 26 in 71 fathoms and going out to the deepest haul of 90.5 fathoms. Large catches of POP dominated the catches in six hauls, 28, 29, 30, 33, 34, and 37, with catches of 1,000 to 5,000 lb.—all are circled in Drawing 8. They were close to the continental slope, where POP were believed to straddle the break at 100 fathoms. *S. brevispinis* was found in every haul in small catches of 20 to 200 lb. POP was found in eleven of the hauls from 71 to 90.5 fathoms, with six of the hauls ranging from 1,000 to 5,000 lb. Fishermen knew POP were found near and on the continental break at 100 fathoms (600 feet). One other species that had a large catch was *S. rubrivinctus,* where haul 33 had 1,000 lb. at a depth of 84.5 fathoms.

About nine years before, during Cruise 9, in the area just south of Area 5, outside the 100-fathom contour on the slope, good catches of 1,000 to 4,000 lb. of POP were taken in three hauls between 125 and 150 fathoms. A lot of time was spent during Cruise 9 mending or replacing the damaged trawls before locating successful hauls. Comparing the successful hauls made during Cruises 9 to 46, thirty-eight possible net drags were determined to be trawlable after using the new sounder and the chain during Cruise 46. When a trawl replaced the chain and the haul repeated, thirty-four of them were successful (90 percent), and four unsuccessful (10 percent). Comparing to Cruise 9, there were a total of 61 hauls made in Cruise 46. Thirty-seven were successful (61 percent), and twenty-four were unsuccessful (39 percent). Using the new sounder and the chain saved a lot of mending time on the trawls.

Table 2 lists the species of rockfish found in the thirty-four successful trawl hauls made during Cruise 46 in 1960. The number of hauls that each species was found in the square next to the species' name. For example, *S. flavidus* was found in sixteen of the thirty-four hauls. Species that were found during Cruise 9, made in 1951 and taken in deeper water on the slope, are marked with an X indicating that they were present. There were eleven species of rockfish taken on the shelf during Cruise 46 in water depths of 57 to 92 fathoms versus five taken on the slope during Cruise 9 at depths of 100 to 225 fathoms. No black rockfish were taken during Cruise 9. Only POP were taken on both the shelf and slope sides. The four species of red rockfish were found only on the slope side.

Table 2 Rockfish Species Taken During Cruises 9 and 46.

Year	1960	1951
Cruise No.	CR # 46	CR # 9
Depth Fathoms	57–92	100–225
No. of Hauls	*34	37
Continental	Shelf	Slope
Black Rockfish		
S. brevispinis	**30	–
S. flavidus	16	–
S. entomelas	–	–
S. melanops	3	–
Red Rockfish		
S. aleutianus	–	X
S. alutus, POP	16	X
S. crameri	–	X
S. diploproa	–	X
S. elongatus	7	–
S. paucispinis	19	–
S. pinniger	27	–
S. proriger	4	–
S. rosaceus	–	X
S. ruberrimus	5	–
S. rubrivinctus	14	–
S. wilsoni	1	–
No. of Species	11	5

*The thirty-four successful trawl hauls made.
**The total number of hauls each species were found in.

Rockfish X = present, – not present

Chapter 4: Endnotes

1. H. E. Crowther, "Exploratory Fishing," *Fishing Gazette*, 1949 Annual Review Number 66, no. 13 (1949), pp. 106.

2. Thomas O. Duncan, "The Pacific Region of the Bureau of Commercial Fisheries," *Fish and Wildlife Circular* 108 (May 1961).

3. Dayton L. Alverson, "The Japanese and Russian Trawl Fishery in the Bering Sea," *Western Fisheries* (April 1960), pp. 12–14, 30–31. MS #53.

4. D. L. Alverson and N. J. Wilimovsky, "Fishery Investigations of the Southeastern Chukchi Sea," in *Environment of the Cape Thompson Region, Alaska*, ed. N. J. Wilimovsky and J. N. Wolfe (Washington, DC: US Atomic Energy Commission, 1966), pp. 843–860. MS #52.

5. C. R. Hitz, H. C. Johnson, and A. T. Pruter, "Bottom Trawling Explorations off the Washington and British Columbia Coasts, May–August 1960," *Commercial Fisheries Review* 23, no. 6 (June 1961). MS #60.

6. Records of the *John N. Cobb*, Historical Ship Files, ID 119654305, Seattle National Archives Office.

CHAPTER FIVE

My First Trip, Cape Scott, British Columbia 1960

When I came to work on July 5, 1960, I thought that I was hired as a biologist but found that I was hired as a Methods and Equipment Specialist, the first step to becoming a Fishery Engineer. I didn't care what they called me; I had the job I wanted. I had inspected the *John N. Cobb*, which was the first thing that I wanted to do. The second was to find out when I would go out on my first research cruise on the *Cobb*. I found that I had been scheduled for the second half of Cruise 47, a bottomfish exploration off Cape Scott, British Columbia. I also learned that I would be going out with Al Pruter, the Chief Scientist (CS).

The *Cobb* had a break between cruises. They had finished Cruise 46, returning on June 24, 1960. The last haul number used was 38, which was a hang up on the Spit. The first haul number for Cruise 47 was 39 and continued to the last haul, 56. They were scheduled to commence Cruise 47on July 18, 1960. The cruise was divided into two parts. The first half was from July 18 to August 5, and the second from August 15 to September 9, 1960. Waters off the coast of British Columbia were part of the bottom-fishing grounds of the US trawl fleet from 1945 until the Magnuson-Stevens Act became law in 1976 and the two-hundred-mile conservation zone was established.[1] After that date, the US trawl fleet was unable to fish in the Canadian waters and were forced out of Queen Charlotte Sound.

Kevin Bailey's book, *The Western Flyer*, is fascinating, as the vessel he wrote about had become famous since it was chartered by Steinbeck.[2] He made a trip up the east side of Baja California, after which he wrote *The Log from the Sea of Cortez*. The *Western Flyer* was built for the California sardine fisheries. When that fisheries collapsed, it was purchased by Dan Luketa in Washington State. He converted her to a trawler, and she was one of the top US boats that fished for POP. The *Western Flyer* was part of the Washington trawl fleet that fished Queen Charlotte Sound in the late '50s and early '60s. Bailey's book states: "When Dan Luketa and the other local fisherman were dragging for perch, they didn't drop their nets just anyplace looking for fish, nor did they trawl on promising signs of fish seen by the sonar. This was because the bottom ground of the outer continental shelf was rocky and full of snags. Instead they fished on established locations, or 'tows.' These were proven trawling stations that were passed from fisherman to fisherman."

The US trawl fishermen reported areas off Cape Scott, at the end of Vancouver Island, untrawlable. They pointed out areas off the tip of Vancouver Island that they wanted the *Cobb* to explore for possible tows. Our job during Cruise 47 was to discover tows within the untrawlable area and determine what species were present. The *Cobb* returned to Seattle in the middle of Cruise 47 for supplies and to change the scientific staff. Then on August 15, 1960, she headed out the Straits of Juan de Fuca to commence the second half with me aboard. We took several Bathythermograph (BT) casts along the outside of Vancouver Island. It would have been more comfortable to take the Inside Passage on the eastern side of Vancouver Island. I should have had a clue of what was coming when the cook replaced the tablecloth on the galley table with a rubber mat and installed the railing around the top outer edge of the table, designed to keep dishes on it and not in our laps when we got offshore.

My first job was to make a BT cast (Fig. 16) because the Halibut Commission asked us to collect the temperature data from a series of stations while we transited the coast. The device is a long tube with fins at one end and is lowered by a cable from the side of the ship. A coated glass slide is placed inside before it is deployed. A needle traces the temperature onto the slide as it descends as well as when it's retrieved. After the BT is retrieved, the slide is removed and stored in a box. It was night, and we were offshore near the Lightship Station *Swiftsure* northeast of Cape Flattery. I was seasick, so when the vessel stopped at the location of the station, I went out on deck to hook up the BT. Al followed to make sure I was stable enough not to fall overboard while launching and retrieving it. I had to load the slide into the BT and attach it to the cable. The skipper started the winch overhead, taking up the slack as I put the BT over the side into the water. He then let it down to the desired depth and hauled it back up, stopping just below the surface. Using a pole with a wire loop on the end, I caught the cable so that when the BT broke the surface, it wouldn't bang against the side of the rolling vessel. Once above the surface, the BT was brought aboard, the cable disconnected and placed in the rack, and the slide retrieved. I accomplished my first task at sea. The vessel got under way again and headed northwest for the next station. Then I took the slide into the vessel and transferred the information recorded on the slide to a form which described the thermocline that was tranced on the slide.

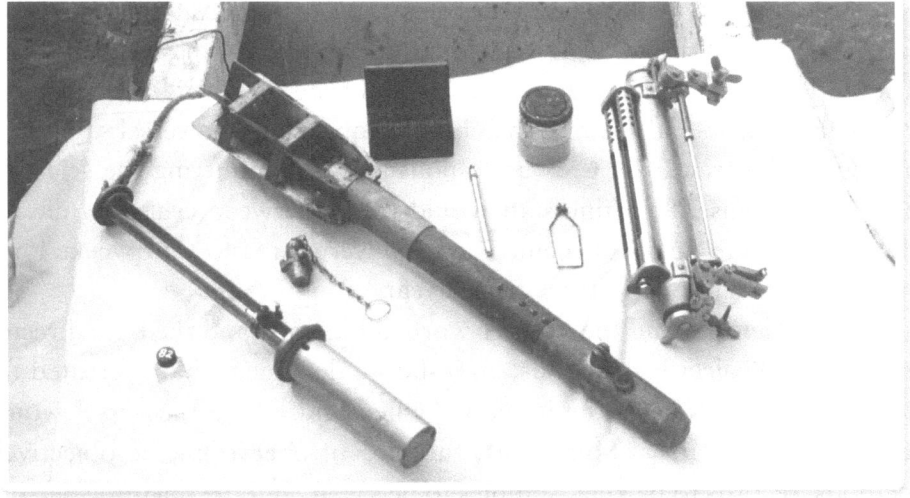

Fig. 16 Bathythermograph (Center of Picture). Image #3302.

The next morning, the *Cobb* kept running northwest as we stopped periodically at specific locations to make BT casts along Vancouver Island. I went to breakfast, but because of the ocean swells, I wasn't feeling too good and hung back. There was a heavy westerly swell all day, and we were quartering them off our port bow. I sat on the seat on the outside corner of the table, where you could look out the galley door, the exit to the open deck, if I had to lose the meal. I kept it down but didn't feel great. Still fighting seasickness, I went outside by the portside door and leaned against the side of the house, looking at the ocean's horizon to the southwest and trying to forget the smell of toast and the taste of chalky milk. It was years before that negative smell was gone from my mind. Later, a crewman told me I had sat in the skipper's place. It was the closest to the bridge so he could respond quickly in case of an emergency. Not a good start for my career.

The afterdeck was wet as seawater came in through the scuppers as the vessel proceeded through the swells and chop toward Cape Scott. I was at sea, aboard the *Cobb* for the first time, and the statement that I heard at the office was going through my mind: Go to sea and publish, and you will be rewarded! My wife, Maureen, was pregnant with our first child, and I'd reached an impasse at school—I could not pass the foreign language exam for a master's degree. Even if seasickness was a problem for me, it wasn't as bad as taking that test again and failing it for the seventh time. My education had already proved its worth. I was employed and aboard the *John N. Cobb*. It was a thrill to be part of this crew—seven fisherman and two scientists, Al and myself. The vessel was a tool used by the scientific staff. The ship's crew was made up of commercial fishermen headed by Captain Pete Larsen, who had the responsibility for operating the vessel safely and managing the crew. The CS was responsible for conducting the scientific part of the cruise. Scientific staff would rotate between cruises while the ship's crew were on all of them. The trips were three to four weeks long, and scientists had two to three trips a year.

Chief scientists and the skipper worked together, and the base director issued both Project Instructions, an in-house document that is created for each cruise which gives the itinerary, when and where the vessel would leave and return to its home port, the area of operation, the objectives, methods, records to be kept, samples to be saved, and the scientific personnel assigned to the trip.

Before I went out on the *Cobb,* the crew aboard her during Cruise 46 developed a system which reduced the damage to the trawls that occurred during Cruises 9 and 13. During the latter cruises, when the echo sounder found enough flat bottom for a haul, the trawl was then towed with the hope that it would be successful and not hang up. The vessel now had a new white line echo sounder installed before the start of Cruise 46, which separated the bottom from the fish and determined whether the bottom was soft or hard. The old sounder that was used before during Cruises 9 and 13 did not have it. With the new sounder, the vessel would run a series of sounding transects to determine the character of the bottom, and once they found enough flat area for a drag, they would drag a chain between the doors, replacing the net. Through trial and error, they found that when the sounder showed soft bottom, it was trawlable, but if hard, it was not. If successful with the chain, they would replace it with a commercial otter-trawl net to evaluate the commercial potential of the groundfish present. For Cruise 47, we had the following types of stations: sounding transects, chain drags, and otter-trawl sets.

Once we arrived on the grounds, the first thing we had to do was duplicate haul 44, made at the end of the first half of the cruise, where the catch was lost before it could be brought aboard. We needed to determine what species were found there. Considerable effort had already been spent in finding this haul and suitable trawling bottom. This was the final stage to determine what species were available and whether a haul could be repeated without losing the catch, as had happened during the first half. The skipper lined up the vessel on a duplicate course along a depth contour of 85 fathoms and gave the command to set the net for haul 45. The description of the East Coast side trawler by Knake is very similar to the *Cobb*, which tows from the stern.[3]

Setting

The setting of the *Cobb's* trawl can be divided into six items, which are outlined in Drawing 9.[4]

Item 1, Arnie, one of the fishermen, heaved the cod end over the stern as the *Cobb* slowly moved forward. The rest of the net followed as the drag of the cod end in the water pulled it off the deck.

Item 2 shows one of two cables, one on each side of the vessel. Each cable was made of three parts: trawl bridle, idler, and trawl cable, all stored on each of two trawl drums of the trawl winch. The idler was attached to the trawl bridle by a jam link, and the other end was attached to the trawl cable by a flat link. As the *Cobb* moved forward, the trawl bridle was played out through a ring in the figure eight. The figure eight was attached to the after trawl door bridle, which hung loosely from the resting trawl door.

Item 3 shows the jam link. Once the jam link came into contact with the figure eight, there was a violent reaction. The net, which was in the water aft of the *Cobb*, all of a sudden became a sea anchor. The trawl bridle took the strain of the net through the trawl doors that were chained to the stanchions. The doors leaped up from a resting position and were held in place by a chain connected to each of the stanchions from the strain of the net, which was trying to stop the vessel's forward movement. Then the main trawl cable and idler became slack. On the opposite side of the door, there was a chain hooked to the stanchion.

Item 4, two fishermen on deck, Arnie on the starboard side and Connie on the port side, connected their individual trawl doors to the main trawl cable by taking the G-hook hanging slack from the door. Sliding the open

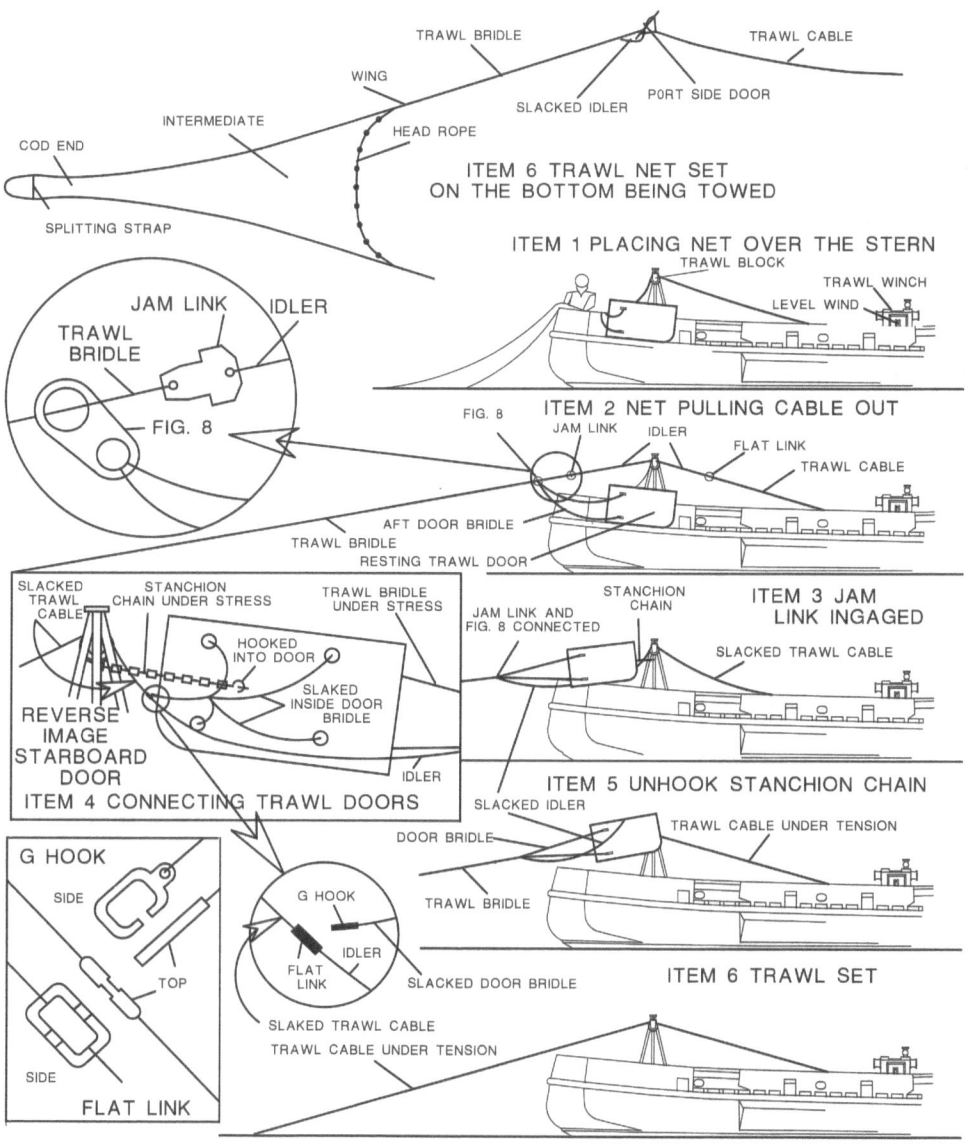

Drawing 9 R/V *John N. Cobb*—Setting the Trawl.

slot on the G-hook through the groove in the flat link which joined the door bridles to the trawl cable. The forward end of the slack idler was looped over the top of the door and attached to the door. The figure eight with the jam link held in the hole with the tension of the net pulling on it with the trawl doors held by the stanchion chain.

Item 5, the winch operators pulled the trawl door up to the block on the trawl stanchion, shifting the pull of the net to the main trawl cable. The chain connecting the trawl door to the stanchion went slack, allowing the hook to be easily released. I had done the same thing on the *Commando*. Forces were great on the trawl doors as the vessel slowly moved through the water. I must admit it was stimulating to connect and disconnect the flat link, hook and unhook the chain from the doors, and be so close to them after seeing them move violently when they took the strain. I swear you could feel the strain of the vibrations of the taut cables. You didn't want to spend much time near the doors when they were absorbing all that energy.

Item 6, the doors were then let out until they were just below the water's surface. The winch brakes were applied, and the doors sheared out to opposite sides of the vessel like kites in the wind. Cable was then let out by each winch operator at the same rate on each side, measured in fathoms. The operators would call out their readings to ensure both sides went down evenly. These procedures reduced the chance of the doors crossing and ensured the net would fish properly.

A sheet of paper, referred to as a scope ratio, in the pilothouse gave the length of trawl cable needed to be let out for a specific water depth. Once the depth was defined for a drag, the table showed the amount of trawl cable needed. When the trawl winch operators reached that amount, the winches were stopped and the brakes set. Then the start of the haul was recorded. The net was then towed along a specific depth contour for at least an hour. Once the winch operators started to retrieve the cable, the haul was recorded as ending.

Retrieving

I still remember the excitement of retrieving the trawl.[5] I had been waiting patiently for the command from the skipper for an hour and a half of towing the trawl along the bottom, hoping it would not hang up on an unknown obstruction and would yield a catch. Once the skipper gave his command, the *Cobb* slowed down, and the winches were engaged. Then the trawl cables were retrieved, and the winch operators used the level wind to make sure the cable was laid evenly on each of the winch drums. The level wind is part of the winch, a device in front of the trawl drum that moves right and left as the cable is retrieved. Once the trawl doors could be seen, the winch operators would bring each door up to the trawl stanchion as far as it would go and then stop.

The next procedure was just the reverse of what occurred when setting the trawl (Drawing 9). The chain hanging loosely from the stanchion was hooked onto the door. The trawl cable was slacked off, and the doors and the chain attached to the stanchion took the strain of the net as the jam link in the figure eight took the force. The G-hook was released from the flat link on the main trawl cable, freeing the doors from the main trawl cable. The doors would hang loosely from their stanchion chain. Then the winches were engaged, putting tension on the idler, which, in turn, freed the jam link and the trawl bridle ran through the hole in the figure eight. The level wind on the winches wound the trawl cable correctly onto the drums followed by the idler and trawl bridles. Once the wings of the net reached the stern, one on each side just aft of each of the trawl stanchions the winches were stopped. The trawl was long and divided into three parts—the wings, the intermediate, and the sausage-shaped cod end.

The wings were held in position by the winch's brakes as the vessel continued moving forward slowly. A V-shaped lifting bridle was obtained from storage and attached to each of the wings. The other end of the bridle was hooked into one of the singles hanging from the block at the end of the boom. Then the wings were raised as far as they could go to the block at the end of the boom. The vessel was turned to starboard and then put into reverse, bringing the net to the starboard side.[6]

The net was lowered, and the intermediate part of the net would fall over the rail as the wings were let down onto the deck. The intermediate part of the net was held in place as it was folded over the rail. The lifting bridle was disconnected from the wings as well as from the single. The single could then be wrapped around the intermediate just outside the rail and lifted to the block on the boom. The net would then be dropped, falling across the rail and held in place. The procedure could be repeated until the catch in the cod end was reached. If the catch could be brought onboard in one lift, the puckering string was pulled, letting the catch fall into the checker.

If the catch was too large to be lifted, it needed to be split by a splitting strap — a line encircling the cod end through rings attached to the webbing. By hooking it with the single line from the boom overhead and pulling it up, the line encircling the cod end through the rings tightened, forcing the catch to split. Part of the catch remained in the cod end, and the rest of it slid to the intermediate part of the net alongside the vessel, where it remained trapped. The cod end was lifted out of the water and released into the checker. The puckering string was retied, the bag put back into the water, and the process repeated until the entire catch was removed from the net.

Sorting

It was exciting to see the fish come aboard and be released into the checker. This was my first opportunity to be involved in sorting the catch while working with the CS and recording it. The first thing the biologist had to do when the catch was brought aboard was make an estimate of the total weight. A maximum lift of a full cod end weighed about 2,000 lb. and would fill the checker when the fish were dumped into it. A half checker was 1,000 lb., a quarter checker 500 lb., and a three-quarter checker 1,500 lb. Al estimated the total weight of this catch was 2,000 lb., or a single lift, which was recorded.

The skipper on the bridge had a chart of the area laid out on the chart table, where he plotted the start and end of the tow, drew a line between the points, and wrote the station number next to it, which in this case was haul 45. He used loran or radar to get the position and depth of the drag from the sounder. When making a tow, he tried to follow a depth contour, which was 85 to 86 fathoms in this case. The biologist later converted the plot into latitude and longitude from the chart and recorded them on the form, along with any other information the form required. The positions were extremely important, so the fishermen who wanted us to explore the area could duplicate the tow. This was exactly what we did on the first tow we made after the run up the west coast of Vancouver Island.

The form we used divided the catch into four different categories: flatfish, round fish, rockfish, and other (scrap fish). The catch was sorted into the four groups. Each group was broken down by individual species. Each species' weight was recorded, and then the total weight of each group was determined. In this case, there was 500 lb. of flatfish, 570 lb. of round fish, 970 lb. of rockfish, and 50 lb. of scrap fish for a grand total of 2,090 lb. This amount was recorded as the official total weight of the catch. It gives the CS a check on the original estimated weight of this haul, which was 2,000 lb.

Once the catch was sorted into the four groups, I was excited to find five different species of rockfish making up the 970 lb. in that category. My excitement grew when I started to recognize one of them as POP. We were required to identify each of these five rockfish and record them by their scientific names, because there were at least twenty-four species of rockfish reported along the British Columbian coast.[7] They were separated on deck into three subgroups: black, POP, and red. In the black group, there was only one species, *S. brevispinis,* with a total weight of 70 lb. Once POP were recognized, identified, and separated from the other red rockfish, their total weight was 500 lb. Of the remaining 400 lb. of red rockfish we found, there were three species: 250 lb. of *S. pinniger*, 140 lb. of *S. rubrivinctus*, and an individual *S. paucispinis* that weighed 10 lb.

I was so focused on the catch, helping work up the data and identify the rockfish that I completely forgot about being seasick on the transit along the outer coast of Vancouver Island. Once the catch was recorded, we moved to another part of the exploratory area north of Triangle Island. This was only the first day on the grounds, but a good preview of the rest of the cruise.

During the first half of the cruise only six hauls—39 to 44—were made with a net in the southern part of the survey (Drawing 10), where the bottom was soft. The chain was used, and if it was successfully towed for at least an hour, the net was tried. Three areas were found to be soft and trawlable. Haul 43 hung up badly, losing the catch, and haul 44 lost the catch when the net malfunctioned. Haul 39, made in 58 fathoms of water, had a catch dominated with 1,300 lb. of turbot, not a valued commercial flatfish. Hauls 40, 41, and 42 were successful, made at water depths of 104 to 116 fathoms at the top of the slope. All three had good catches of 1,000 lb. or more of POP. Haul 40 was dominated by four species: Dover sole (1,200 lb.), turbot (1,700 lb.), *S. brevispinis* (1,300 lb.), and POP (1,000 lb.).

I was interested in the rockfish catches that occurred in the five successful hauls during the first half of the cruise (Appendix 5). There were three species of black rockfish. *S. brevispinis* was found in all five hauls, with haul 40 producing 1,300 lb. *S. flavidus* was found in three hauls and *S. entomelas* in two, both in small quantities. There were seven species of red rockfish. POP was taken in four of the hauls, with hauls 40, 41, and 42 ranging from 1,000 to 1,650 lb. and taken on the top of the slope in waters ranging from 104.5 to 115 fathoms. The other six species of red rockfish—*S. elongatus*, *S. paucispinis*, *S.pinniger*, *S. rubrivinctus*, *S. saxicola*, and *S. zacentrus*—were found in one to three of the hauls and in small quantities of a trace to 250 lb.

Eleven Net Hauls

When the *Cobb* returned with me aboard after the cruise break, she duplicated haul 44 with haul 45 to determine what species of fish were present. After completing haul 45, the *Cobb* moved to just northeast of Triangle Island in the Scott Island group. The skipper had a navigational chart of the area laid out on the chart table. We recorded the movement of the vessel across the area by drawing course lines on the chart. The type of bottom the sounder showed, either soft or hard, was added to the chart. Once the chart showed enough soft bottom where a tow of an hour or more could be made, the net was replaced with a long chain so the doors could spread normally and the chain could slide along the bottom. If it encountered an object and hung up the chain, the tow would be abandoned. If it was clear, the chain would be replaced with the net, and the tow repeated. If successful for at least an hour without hanging up, the catch would be recorded. Eleven net hauls were tried, 46 to 56; ten of them were successful and one was not, haul 56, which hung up (Drawing 10). All were made in the confined soft bottom area in the northern part of the surveyed area. All the successful hauls lasted an hour or more and were made along a different depth contour, starting at about 51 fathoms of water and reaching 113 fathoms. The continental shelf ends at 100 fathoms, and the slope begins, being defined as the continental break.

I helped fill out the exploratory catch records for each haul, the same form we had just used for haul 45. The rockfish group was subdivided into black, POP, and red. It was important to split POP out from all the other rockfish that could be taken along the Pacific coast. The market for POP was growing on the East Coast, and the US Army was issuing contracts for POP fillets, which were ideal for IQF (individual quick freezing). The Pacific fish houses could buy large quantities of POP, but they could only buy small quantities of other rockfish that they could sell. Therefore, the fishermen were required to separate the rockfish out at sea.

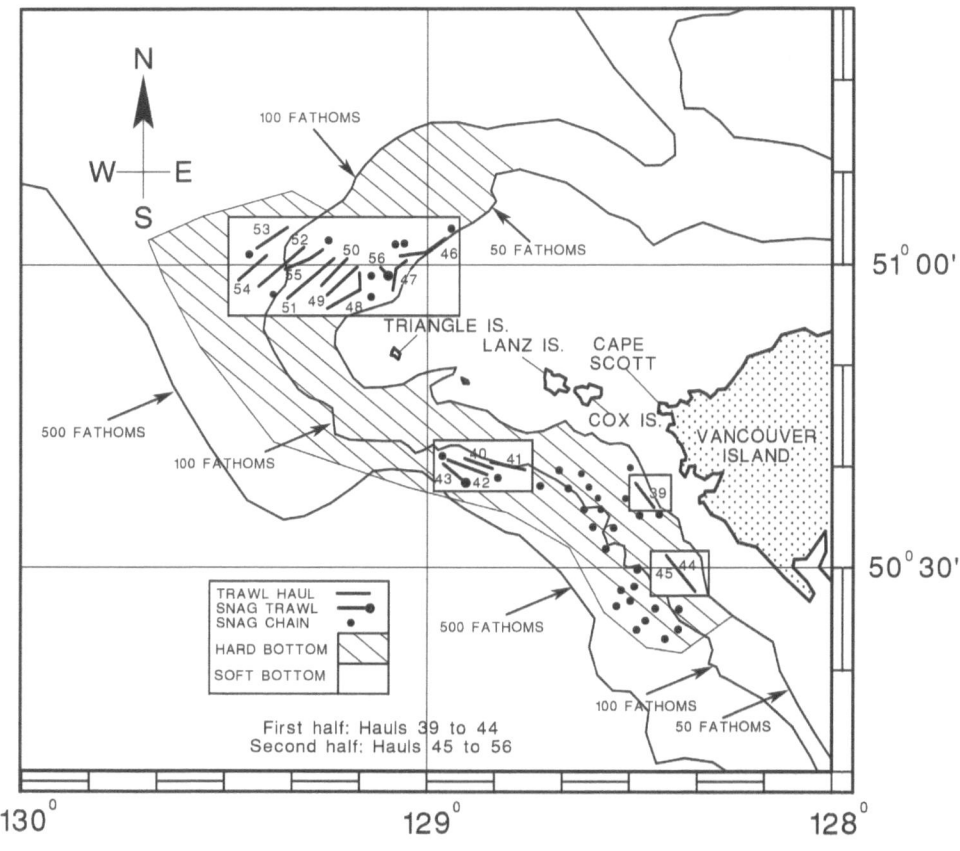

Drawing 10 R/V *John N. Cobb*—Cruise 47 Successful Hauls.

The catch of rockfish was the first time I could see the changes that occurred in the species composition as we moved to deeper water. I had no prior knowledge of the past cruise information. I had just come from the College of Fisheries and had some knowledge about the POP from my summer job when I worked at a fish house on the waterfront, placing POP fillets on a light table (candling) and counting the number of blemishes in the fillets.[8]

The rockfish catches taken in the series of trawl catches were arranged by chance to go from shallowest depth on the shelf to the deepest on the slope (Appendix 6). Starting with haul 46 at 51 fathoms on the shelf, we moved deeper with each station out to 113 fathoms on the slope. Haul 55, which was calculated at an average depth of 96 fathoms, was moved to the last haul on the shelf followed by 52, 53, and 54 on the slope. Each haul was towed along a depth contour for about one hour. The squares in the appendix where the catch was 1,000 lb. or more are highlighted by wider lines surrounding the square. The continental break is also noted at 100 fathoms with a heavy black vertical line.

The large catches of *S. brevispinis,* a black rockfish that occurred on the shelf, and the large catches of three species of red rockfish—POP, *S. piniger*, and *S. ruberrimus*—gave me the feeling that the rockfish changed with depth. No rockfish were found in the first two hauls, 46 and 47. Two species of black rockfish were easily identified. *S. brevispinis* was found in all eight hauls, with the largest catches occurring on the continental shelf—haul 48 had a catch of 3,500 lb.; haul 50, 1,500 lb.; and haul 51, 2,500 lb. *S. flavidus* was taken in four hauls scattered throughout the depth range in very small numbers of 20 to 50 lb. In the red rockfish category, there were seven species. *S. alutus* (POP) was first found in haul 50 in 83 fathoms of water. Two hauls had large catches of POP: haul 55 made in 96 fathoms with a catch of 1,000 lb., and haul 53 made at 112.5 fathoms on top of the slope with a catch of 1,100 lb. *S. elongates* were small and easily identified by the four green stripes on the side of the body. They were found in shallower depths of 70 to 85 fathoms in hauls 48, 49, and 50 in small numbers. *S. paucispinis* was found in four of the hauls in small amounts. They could easily be separated from POP and the rest of the rockfish by their large size, greatly protruding lower jaw, and lack of a symphyseal knob.

A problem became apparent when a species of red rockfish dominated the catch and was mixed with POP, as occurred in hauls 51 and 52, where *S. pinniger* dominated the catch in quantities of 1,500 and 1,100 lb., whereas there were only 400 and 800 lb. of POP. Separation could be made by *S. pinniger's* larger size, color, slant of the anal fin, and the lack of the symphyseal knob. The one I found most difficult to separate from POP was the *S. proriger*, found in small catches in hauls 51 and 55. They were about the size of POP and had a similar symphyseal knob and dark spots on the back, the keys for identifying POP, but the red stripe along the lateral line gave them away. The first species that was recognized because of its large size and color was *S. ruberrimus*, often referred to as red snapper or buoy keg by the halibut fishermen. It was found in all eight of the ten hauls rockfish were taken in, with the largest catch of 3,000 lb. in haul 54, made in water depths of 111 to 115 fathoms. The last species in the catch of rockfish was *S. saxicola,* which occurred in three of the deeper hauls. Their color and markings were similar to POP, blotches along the back were similar and were difficult to separate from POP, but as the common name implies, the green streaks on the tail and smaller size distinguished them from POP. With Al's help and the existing keys, we identified each species of rockfish.

Looking at the catch of rockfish from all of Cruise 47, I began to agree with Lee Alverson that POP would become a dominant commercial fisheries because of the three large catches of them on the slope side, two taken during the first half of the cruise and one from the second half. If there was a larger market for *S. brevispinis,* it could potentially be a dominant commercial rockfish species.

Table 3 lists the total rockfish taken in the fifteen hauls made during Cruise 47. There were ten hauls made on the shelf side and five hauls on the slope. POP were found mainly on the slope side and in only ten of the hauls, which brought some doubt. Whereas *S. brevispinis* were found in thirteen of the hauls, seven on the shelf and six on the slope. There were twelve different rockfish species encountered in the fifteen hauls made during the trip. One species of interest, *S. pinniger,* was found in eight hauls, with hauls 51 and 52 producing large catches that straddled the continental bridge. I was anxious to get back from the trip and learn more about POP and looked forward to going out on my second exploration.

Sportfishing

One evening, we anchored in a bay just south of Cape Scott and another time alongside Triangle Island, giving the crew a chance to sportfish. I was struck by the variety of rockfish caught. I still remember seeing my first *S. nebulosus*—its blue-black body has a vivid yellow strip down the side—and was surprised to see *S. caurinus,* with which I was very familiar since it was part of my studies at the College of Fisheries. I had caught *S. melanops* in the San Juan Islands, but *S. maliger,* with its varied pattern of brown, was new to me. Two species the crew took with the pole and line sports gear were the *S. pinniger* and *S. ruberrimus,* both caught in the trawls during this trip. The intensity of the different colors displayed by these rockfish when they came out of the water was amazing. When we used the trawl net in much deeper waters, the colors were not as vivid as the sport-caught fish but were still noticeable.

By the time I was hired, we were required to identify each species of rockfish in the catch. In those days, the identification keys were put

Table 3 Rockfish Species Taken During Cruise 47.

Year	1960	1960	1960
Cruise No.	CR # 47	CR # 47	CR # 47
Depth Fathoms	58–115	58–100	100–115
No. of Hauls	*15	10	5
Continental	Both	Shelf	Slope
Black Rockfish			
S. brevispinis	**13	7	6
S. entomelas	2	–	2
S. flavidus	7	3	4
S. melanops	–	–	–
Red Rockfish			
S. aleutianus	–	–	–
S. alutus, POP	10	4	6
S. crameri	–	–	–
S. diploproa	–	–	–
S. elongatus	4	3	1
S. paucispinis	7	5	2
S. pinniger	8	5	3
S. proriger	2	2	–
S. rosaceus	–	–	–
S. ruberrimus	8	5	3
S. rubrivinctus	2	1	1
S. saxicola	5	1	4
S. wilsoni	–	–	–
S. zacentrus	3	–	3
No. of Species	12	10	11

*The fifteen successful trawl hauls made during the cruise.

**The total number of hauls each species were found in the fifteen.

together in the depository where the preserved specimens were kept, and measurements and counts were made on individual specimens. Most were preserved in formaldehyde, which leached out the natural color. The keys were based on different characteristics of individual species preserved in the museums and were published. We used Clemens and Wilby's *Fishes of the Pacific Coast of Canada* and Phillips's *A Review of the Rockfishes of California*.[9] In 2008, after the *Cobb* was decommissioned, I found a copy of the Clemens and Wilby book we had used. On the inside of the cover, I found "Exploratory Fishing—US Fish & Wildlife" written in my handwriting when I was on a trip in 1960. Both keys used the true-false method for identifying a fish, such as the statement, has gill cover—yes or no. If you said yes, you went to the next statement, until you reached the species you wanted to identify.

If we couldn't identify the rockfish based on the Clemens and Wilby key, which listed twenty-four species of rockfish in Canadian waters, we would turn to Phillips's key, which lists forty-nine species. His key was printed in even smaller font. You can imagine trying to identify a fish by following a key like this on an open deck in the rain and wind, the vessel rolling and pitching from ocean swells and chop. Inspecting them while wearing bulky foul weather gear and trying to read fine print, answering true or false questions, and finally, reaching the fish in question was extremely difficult. Keeping the book dry was almost impossible. If you got out of the weather and went down into the hold where the scientific lab was located, I, for one, would become seasick and have to get back up on deck.

With so many different rockfish, we needed a better field key to identify them correctly. I thought there had to be a better way for identifying rockfish in the field. Color photos of each species would have been the ideal way to produce a field key, but the cost of printing a colored pamphlet in those days was prohibitive, and there weren't many colored photos of rockfish available at the time.

When the weather became too rough to fish, we headed inshore to resupply and pick up the mail at Port Hardy. One day, the skipper's log indicated there was a heavy westerly chop and swell with a WSW wind around 30 knots and a sea height of about 13 feet. It was on our stern as we headed for shore, and it was invigorating. I was able to stand in a protected space under the eaves of the pilothouse on the bridge and could watch the wind-driven waves going in the same direction as us. I could feel the stern rise and the vessel pick up speed as she was shoved forward by a wave then settle as she crossed the trough, when the bow would rise and we would slow down as the wave broke and the white water foamed around the bow. The stern would lift again, repeating the process over and over. She was running before the wind and swells. I really understood the term "Fair winds and following sea." My seasickness had vanished for the moment. Thank goodness we weren't going the other way.

We departed the area on September 6, 1960, and retook the same eleven BT stations we had worked on the way up along the outer coast of Vancouver Island. It was a better trip back because we had a following sea, and I apparently had my sea legs. Our last BT station was next to the Lightship Station *Swiftsure*, the same one I had taken thirty-four days before, off the Strait of Juan de Fuca. We stopped at Neah Bay and cleared Customs, then it was into the strait again, heading in the right direction, east toward home. Railings on the galley table were removed, and the rubber mat on the table was replaced with a white tablecloth. Finally, we were in calm waters and on our way home! We stopped in Port Angeles and cleaned up the vessel. I took the opportunity to call home from a phone booth on the dock. The next morning, as we neared Point No Point and could see the locks, it seemed as if we were standing still and would never reach them. The *Cobb* was so slow. We finally arrived at our moorage at the Stoneway dock on September 9, 1960, completing the cruise.

Chapter 5: Endnotes

1. Fred C. Cleaver, *The Washington Otter Trawl Fishery with Reference to the Petrale Sole* (N.p.: State of Washington, Department of Fisheries, April 1949).

2. Kevin M. Bailey, *The Western Flyer* (Chicago: University of Chicago Press, 2015).

3. Boris O. Knake, "Operation of North Atlantic Type Otter Trawl Gear," *Fishery Leaflet* 445 (May 1958).

4. Charles R. Hitz, "Setting the Trawl-My First Cobb Trip," Carmel Finley (blog), September 2, 2014. https://carmelfinley.wordpress.com/2015/08/11/setting-the-trawl-my-first-cobb-trip/. Bob's posting #33.

5. Charles R. Hitz, "Retrieving the Trawl-My First Cobb Trip," Carmel Finley (blog), January 2, 2016. https://carmelfinley.wordpress.com/2016/01/02/retrieving-the-trawl-my-first-cobb-trip/. Bob's posting #37.

6. James A. Cole, *Drawing on Our History: Fishing Vessels of the Pacific Northwest and Alaska* (Seattle: Documentary Media LLC, 2013), pp. 163.

7. W. A. Clemens and G. V. Wilby, "Fishes of the Pacific Coast of Canada," *Fisheries Research Board of Canada Bulletin* No. 68 (1961).

8. John Liston and Charles R. Hitz, "Second Survey of the Occurrence of Parasites and Blemishes in Pacific Ocean Perch, *Sebastodes alutus*, May-June 1959," *Special Scientific Report—Fisheries,* No. 383 (June 1961).

9. J. B. Phillips, "A Review of the Rockfishes of California (Family Scorpaenidae)," *California Fish and Game Fish Bulletin*, No. 205 (1957).

CHAPTER SIX

My Second Trip, Heceta Bank, Oregon 1961

On the morning of April 25, 1961, I was aboard the *John N. Cobb* on the first half of Exploratory Cruise 50, heading for an area off the Oregon coast as part of the scientific party. It was my second cruise, and we were headed toward the same area where Lee Alverson identified the longjaw rockfish (*S. alutus*) in the summer of 1949 when he was on the *Harold A*.

We had left Seattle the day before and were headed south along the Washington coast for an area near Newport, Oregon, where the fishermen wanted us to explore. The weather wasn't perfect in my mind, while others thought it was great for traveling since we were running with it. The vessel was moving south at about 9.5 knots, her normal cruising speed. There was a large northwesterly swell. I stood under the eaves of the pilothouse on the port side, protected from the wind, my favorite place when we were running. I had plenty of fresh air and could see the horizon. Being outside made me more comfortable and kept seasickness under control. It was "half-pill day" weather for me, since Bonamine helped me with my ongoing problem of seasickness and finding my sea legs.

Loran, a navigational device that came out during WWII, was an important aspect of our research. On the way out, we checked our loran set to make sure of its accuracy by checking the location of Lightship Stations *Umatilla Reef* and *Columbia River* when we passed them. On the

Cobb, there was a receiver just aft of the chart table, which would pick up two different loran readings for our location. Comparing them to loran lines on the navigational chart would determine our exact location. As we had done off Canada during my first cruise, we sounded out the areas that fishermen defined as untrawlable. The areas they wanted us to survey were Stonewall Bank and Hecate Bank, located off the coast of Oregon between Newport and the Siuslaw River. Once we arrived there, we began sounding the bottom right away to determine which grounds were soft and which were hard bottom. Through the next two nights and days, the bottom was sounded to determine trawlable grounds.

During this time, I asked the mate why the drags were always made with the bottom depth contour—why not make them from shallow to deeper or deeper to shallow water? He informed me with a smile and the patience of one dealing with a two-year-old that the cable would be retrieved or let out as the vessel changed depth so the net would fish properly. Different species of fish sought a certain depth of water, and it was necessary to seek the depth at which you wanted to tow the net in order to catch the fish you wanted. I should have known. Why did I continue to ask dumb questions?

When the soft bottom areas were defined, we used the chain instead of the net to determine whether they were clear of snags. If so, we could replace the net to see what fish were there. We made several successful tows, but others hung up, stopping the vessel, or snagged, tearing the net and losing the catch. Some of the hang-ups would be so subtle the vessel felt like it was still moving, though it was anchored. We would check the outboard outlet on the port side from the wing of the bridge and check the stream of cooling water from the engine room to see if there was a wake as the vessel was moved forward. If there was no wake, we were anchored. The order would be given to haul in the gear, and the winches would start taking in cable. If the net remained fixed to the bottom of the ocean, the vessel would be pulled back until the net was free.

The skipper kept an eye on the sky, looking for a change in the cloud cover, and would monitor the Coast Guard radio weather broadcasts. If gale warnings were called, he would take appropriate action. Along the Oregon coast, all the harbors have shallow water across them, referred to as a bar, and they become treacherous in bad weather. When waves get too dangerous to cross, the Coast Guard shuts down the entrance to the harbor. The only way to stay safe is to go north to the Strait of Juan de Fuca, miles away, or ride it out. During this study, the closest place for us to gain protection during a storm was to cross the Newport Bar. The best time to cross a bar is at high slack tide. You have more water under the keel and less current to deal with, which reduces the height of the incoming swells. There are two breakwaters on each side of the canal on the Newport Bar and a turn toward the north while coming in from the southwest. Once you make that turn, you come into quiet waters as you go under the main highway bridge.

I have great respect for those bars and am always anxious about crossing them. They are dangerous, and vessels and crews have been lost on them. In fact, one crossing we made on the *Cobb* into Newport was scary. An abnormal swell picked up the *Cobb's* stern as she was making the turn to the north while following the channel into Newport. It pushed us so that we were pointing the wrong way, heading for the northwest breakwater. A small skiff was fishing near it. Our skipper put the helm hard over to starboard, trying to avoid the skiff and the jetty's wall. She finally responded, turning away from the jetty and the skiff, which was also trying to get out of our way. We finally got back into the channel without hitting them, a very close call for both of us. We were all relieved as we went under the bridge into Newport.

Once the weather moderated, we continued our research, looking for trawlable grounds. At night, we would often shut down and drift. For example, one night, we were twenty miles off the Siuslaw River. The main engine was shut down, and only one auxiliary was running, giving the vessel electric power, so the navigational lights kept burning. The skipper tied the steering wheel so the rudder would not work with the swells, which was an ongoing practice at the time. I believe this practice ended when a fishing vessel drifting off the Columbia River with the crew asleep was rammed by another one heading toward the Columbia River. After that, our skipper ensured his vessel was under command at all times. The skipper, mate, chief, and first assistant engineer would each stand their twelve-hour watch, ensuring the vessel was under command twenty-four seven. That procedure was followed through the years until she was decommissioned. I slept much better with a ship's officer monitoring the radar, making sure the *Cobb* would not be the next victim.

When the cruise ended, I was assigned the job of helping write the end report for cruise 50. Like the final published report for Cruises 46 and 47, the Otter Trawl Fishing Log was only included in the reprints of Separate No. 677 of the published document. I again went back to the Archives and recorded the individual species name of each rockfish taken in each net haul. A total of twenty-nine net hauls were made on the grounds the sounder showed to have soft bottom during the first half of the cruise. Eight of them were not successful, making a total of twenty-one successful hauls. They formed four areas (Drawing 11). Area 1 was located west of Stonewall Bank, Area 2 was east of Heceta Bank, and Areas 3 and 4 were along the 100 fathoms contour west of Heceta Bank.[1]

The shallowest part in Area 1 was 66 to 100 fathoms, where hauls 1 through 6 were made on ground the sounder indicated was soft but was not as trawlable as indicated. Only hauls 1 and 5 were successful, with small catches of 270 lb. and 500 lb. of mixed species. The other four were unsuccessful. Hauls 2 and 4 hung up, while hauls 3 and 6 snagged, tearing the net and losing most of the catch.

The *Cobb* then moved south to Area 2, on the shelf in shallower waters of 58 to 80 fathoms of water, where all five hauls, 7 through 11, were made on bottom the sounder showed to be soft and trawlable. But again, it wasn't true since all, with the exception of haul 10, were unsuccessful. Haul 10 had a total catch of 120 lb., with no POP and a trace of red rockfish, a 6-lb. *S. paucispinis*. The other four hauls snagged.

She moved back north to Area 1 again and made seven hauls, 12 to 18, on grounds the sounder indicated were soft, and all were successful this time. They, along with hauls 1, 5 and 19, were entered into Appendix 8. Haul 19 was included since the vessel departed Area 1 for Area 2, making haul 19 on the way. The ten hauls are arranged by depth, the shallowest of 78.5 fathoms to the deepest of 183 fathoms. The rockfish catches by individual species are divided into black and red rockfish. The black rockfish composed three species—*S. brevispinis*, *S. flavidus*, and a new one, *S. entomelas*. All were close to the break at 100 fathoms. Eight species of red rockfish comprised the rest of the rockfish catches. POP were found in nine hauls with small catches from a trace to 944 lb. from a depth of 65.5 fathoms to the deepest haul of 183 fathoms. *S. paucispinis*, *S. pinniger*, *S. elongatus S. rubrivinctus,* and *S. saxicola* were taken in small quantiles. The remaining two red rockfish species, *S. diploproa* and *S. crameri*, I had never seen before. They were found in five of the deepest hauls at 149.5 to 183 fathoms. Haul 13 had a large catch of 1,053 lb. of *S. diploproa*. They were easily identified because of the concave upper jaw. *S. crameri* was identified by the five black blotches on the back. The lack of POP catches of 1,000 lb. or more made off Oregon, especially in the waters deeper than

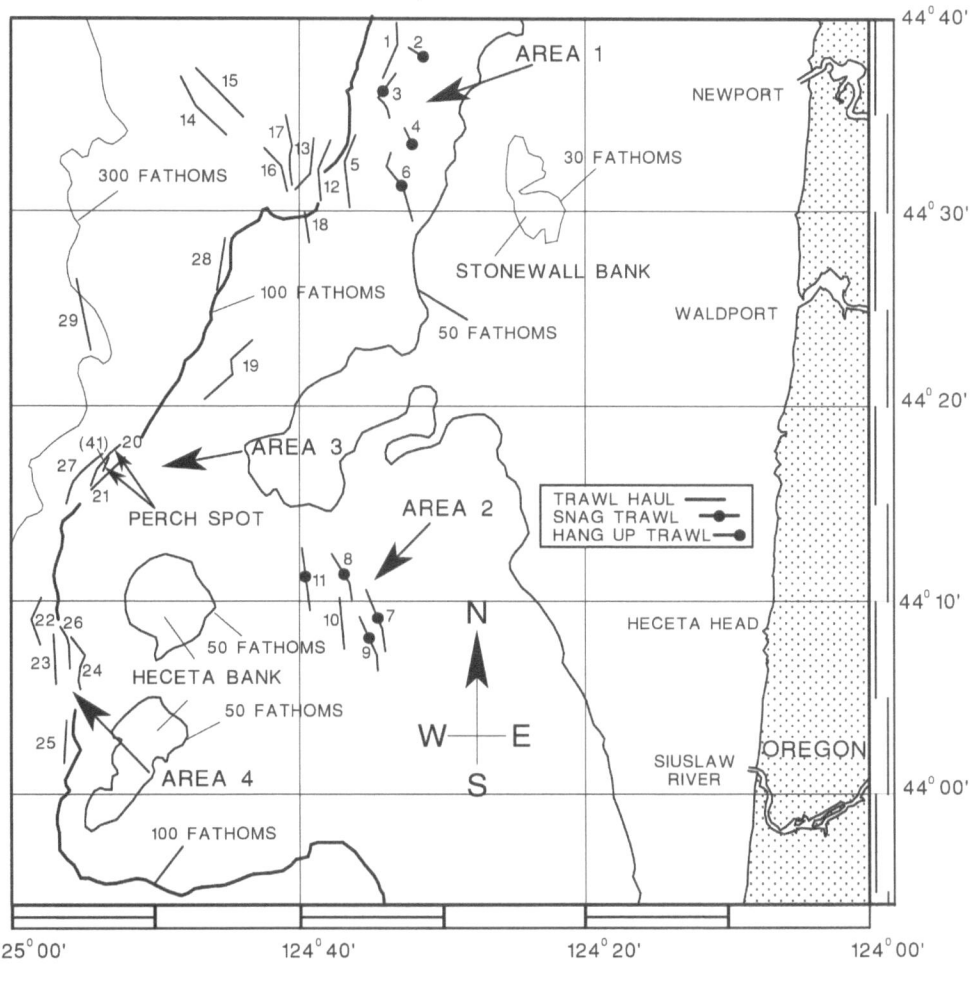

Drawing 11 R/V *John N. Cobb*—Cruise 50 (First Half) Successful Hauls.

100 fathoms, caused me to doubt Lee's idea that POP was the dominant species along the northeastern Pacific coast.

I was surprised to see my first "idiot" in haul 12, a bright red, small, narrow fish with a large head and eyes that differed from the rockfish but were grouped in the red rockfish category. I didn't see them in Canada, but in Oregon, they were found in most of the hauls on the slope. The fishermen along the coast called them idiots, as did the crew on the *Cobb*. It's not a true rockfish, just looks like one. Its common name is the shortspine thornyhead, and its scientific name, *Sebastolobus alascanus*.[2] They are in a different genera from the large group of rockfish found along the coast of the North Pacific. It was first taken in haul 12 at depths of about 120 fathoms and was found in the rest of the deeper hauls out to 183 fathoms in haul 14 (Appendix 7). Joe Dunatov, the bos'n or lead fisherman, became excited when he saw the idiots. He collected several of them for his home pack, which I couldn't believe since they were all head and very little body. The only other time I saw him really get excited was when he spotted an octopus in the catch. I have never seen a person move so quickly—he got it, turned it inside out, and had it drying in the rigging in an instant. I asked his son Tom, years later, after we became good friends, why his dad liked idiots. Tom asked me if I'd ever heard of the wedding fish. The Japanese consider it a delicacy, and it is served during a wedding ceremony. His family loved it.

We moved from Area 1 and headed south to Area 3 along the 100-fathom contour, which we had scouted out with the sounder before. On May 13, we made the first set in Area 3, haul 20. My doubts about Lee's idea vanished in an instant. I saw an unbelievable sight that will remain with me forever. The trawl had skimmed along the ocean bottom 125 to 140 fathoms below the surface. After an hour of towing, I was standing on the upper deck on the starboard side of the after end of the pilothouse, looking aft when the trawl doors reached the vessel. I anticipated that the cod end would float up to the surface to reveal the catch of rockfish. Instead, the ocean behind the vessel erupted violently as a huge ball broke the surface and rose 8 to 10 feet in the air as sheets of water cascaded off it. It happened so quickly I couldn't believe my eyes. The net was full of POP. Their air bladders had expanded as they were brought up from over 100 fathoms of water, forming the cod end into an enormous ball. The air bladder in a rockfish controls its equilibrium at depth, similar to a submarine's method of taking on water to sink and forcing it out with compressed air to rise. The fish's air bladder adjusts internally, so once the fish are netted and raised rapidly through the water with no opening to let the air out, the bladders expand like balloons.

The catch was brought aboard in five splits (2,000 lb. each) and a lift, which brought the total catch to 11,765 lb., almost six tons of fish. The dominant species was POP, with 11,000 lb. or thirty-three thousand individual fish, accounting for 94 percent of the catch. This was the largest catch I had seen. The doubt about Lee's idea was gone. Apparently, when very large catches of POP were taken in past explorations, they were considered pure POP because not enough splits were taken aboard to determine the makeup of the catch. In this case, we brought the entire catch aboard and found that it was not so pure. There were 100 lb. of sablefish, 100 lb. of turbot, 50 lb. of hake, 50 lb. of rex sole, and 40 lb. of Dover sole. There were 125 lb. of black rockfish and 300 lb. of one species of red rockfish, *S. saxicola*, and trace amounts of other species, such as dogfish, ratfish, and skates. There were 765 lb. of fish other than POP, or 6 percent of the total catch.

When haul 20 came aboard and was sorted, we took a subsample of the POP. Each individual was opened to determine sex as well as the

Fig. 17 R/V *John N. Cobb*—Large Catch of Pacific Ocean Perch. Image #11945.

maturity of the females. Individuals were weighed and ranged from 1 to 4 lb., averaging 3 lb. each. Total length ranged from thirty to fifty centimeters (11.7 inches to 19.5 inches), averaging forty-two centimeters (16.38 inches). Otoliths, or ear bones, about half an inch in length, were taken from certain fish by cutting the top of the neck just behind the head and bending it down to expose two otoliths. We then picked each one out with tweezers and put them into an envelope, identifying them with individual fish. They would be given to a biologist who specialized in aging fish. I was interested in the maturity of the females and found all the POP to be in a transitional stage of maturity in May, which would coincide with past findings of them releasing their young in the winter.

On the same day haul 20 was made, haul 21 was made in the same area but at a shallower depth interval, 99 to 101 fathoms. The total catch was 4,100 lb. and was dominated by 1,400 lb. of turbot and 1,055 lb. of *S. pinniger*, with only 292 lb. of POP. Haul 27, made two days later next to both of these hauls but in deeper water of 155 to 172 fathoms, had a small catch of 520 lb., with only 50 lb. of POP.

Appendix 8 lists the ten hauls made in Areas 3 and 4, along with hauls 28 and 29 made north of them. They are listed by depth, shallowest to deepest, with the catch broken down by species. Black rockfish had a poor showing. *S. brevispinis* was found in the first four hauls near the continental break in catches from a trace to 260 lb. *S. entomelas* was found in two hauls in small catches of 25 to 168 lb. There were eight species of red rockfish caught in the ten hauls. POP had a very good showing and were found in seven of the ten hauls, with the largest catches showing in four hauls between depths of 132.5 to 138.5 fathoms and one of them, haul 20, with a mammoth catch of 11,000 lb. The other three, hauls 23, 28, and 25, had large catches of POP between 1,000 and 2,000 lb. Two other species of red rockfish had large catches. *S. saxicola* had a large catch of 1,200 lb. in haul 26 at a depth of 113.5 fathoms. Haul 26 was one of three hauls in water depths of 100 to 132.5 fathoms of water just over the break. The other red rockfish, *S. pinniger,* whose single catch was 1,055 lb. in haul 21, was taken at 100 fathoms of water. *S. rubrivinctus* was found in four of the ten hauls in quantities of a trace to 200 lb. in water depths of 113.5 to 163.5 fathoms. Two other species, *S. elongatus* and *S. paucispinis*, were found in small quantities each in two of the ten hauls. The remaining two species were found in deeper waters. *S. diploproa* was found in hauls 22, 25, and 27, with catches of 190 to 500 lb. in water depths of 138.5 to 176 fathoms. The other species, *S. crameri*, was found in hauls 23, 25, 26, and 27 from a depth of 113.5 to 163.5 fathoms in catches of 100 to 600 lb.

We moved north between Areas 1 and 3 and made hauls 28 and 29, included in Appendix 8. Haul 29 was the deepest haul made, near the 300-fathom contour, with 540 lb. of sablefish, 30 lb. of idiots, and no rockfish in the catch. That was the last haul made for this part of the cruise, and the *Cobb* returned to Seattle on May 17, 1961, completing the first half.

Second Half of Cruise 50

Scientific staff were exchanged and the vessel resupplied for the second half of the cruise, which began May 22, 1961. The objective of the second half of the cruise differed from the first half. This part of the trip was a cooperative trip, with State of Oregon biologists tagging Dover sole along with experimenting with tagging POP. Oregon had a current study of tagging commercial species of flatfish, and Dover sole was one of the flatfish species that lived in the deeper water that the state wanted to tag. The Dover sole was the same target that the skipper of the *Harold A.* was seeking when Lee Alverson came aboard his vessel in 1949 to tag Dover sole and found POP. The current drags that the *Cobb* had discovered during the first half of the cruise would be duplicated along the continental slope of 100 fathoms and deeper. The tows would be much shorter aiding to the survival of the Dover sole. The problem with POP is that they have an air bladder, which the Dover sole does not. The POP air bladders would expand when brought up from 100 fathoms of water. A cut was needed to deflate the air bladder, so the tagged POP would sink back down to the bottom. We hoped they would mend quickly and survive the ordeal. Thirty hauls were made, the majority in Area 1 west of Stonewall Bank, for half the length of time, about twenty minutes. The Oregon State biologists tagged 5,429 Dover sole. They also tagged 175 POP.

On May 28, 1961, two weeks after the *Cobb* made the 11,000 lb. catch in haul 20, Al decided to try it again. He wanted to check whether POP were still there and gain a sample to tag. The tow was only twenty minutes long to keep the catch small. The *Cobb* made haul 41 along 109 to 111 fathoms, which was shallower than haul 20's 125 to 140 fathoms. Again, the ocean exploded, the net rising out and literally blowing apart, leaving POP floating on the surface. The catch was estimated at 40,000 lb. or twenty tons of fish, the largest catch the *Cobb* had ever made at that point.

I was not on this part of the cruise, but Al Pruter was the CS for that second half and kept a personal log as part of the cruise data.[3] In them, I was fortunate to find where he gave details on the calls he made to the fishermen after observing the net exploding in the area. He referred to it as the perch spot. Al and Lee had visited many commercial trawl fishermen beforehand to determine where the exploration should be conducted. Al called several trawlers in the area to let them know of the POP catches the *Cobb* had just made on Sunday, May 28, 1961. Among them were three trawlers working nearby, the *Ruth Ellen*, *Washington*, and *Trego*. On late Monday, he went back to check it and found that a different vessel, the *Madeleine J.*, had just fished the spot and had a floater of POP alongside.

They checked the perch spot again on Tuesday and saw the *Kiska*, *Dave II*, *New Mexico*, and *Oregonian* fishing on known grounds just north of there. Talking via marine radio, the *Kiska* reported they had fished the spot that morning, but that small sablefish had set in. They said some of the boats had worked the spot successfully, but others snagged, tearing the net. The *Kiska* gave them information about a snag in 116 fathoms in the spot area, so the *Cobb* made haul 49 in 117 to 118 fathoms and snagged during the tow, ripping the net badly. However, there were still 1,500 lb. of POP remaining in the net.

On Friday, June 2, 1961, on the way back into Astoria to drop off the Oregon biologists, the skipper of the *Destiny,* Gordon White, called the *Cobb* on the radio to thank them for the good work, saying he hoped the *Cobb* would be able to do more work there in the future. Cliff Hall on the *Pacific Queen* compared fathometers and loran readings with the *Cobb,* saying he was going to try some of the *Cobb* drags.

AEC Track Line

The Oregon biologists got off, and Lee Alverson came aboard. He and Al Pruter set off to establish a track line off the Columbia River for the US Atomic Energy Comission (AEC). They took the last few days of the cruise to use the *John N. Cobb* to lay out a series of stations off the Columbia River, the first of a series of AEC cruises conducted over two or three years. The plan was to establish stations 25 fathoms apart out to 1,000 fathoms, starting at 50 fathoms. This was the start of a joint effort with the AEC. On June 6, 1961, the *Cobb* started looking for a 100-fathom station since the two shallower stations, 50 and 75 fathoms, could be done later when they would be on the continental shelf and on known trawling grounds. They started at 100 fathoms and proceeded out at 25-fathom intervals. During the next six days, they scouted out fourteen hauls for possible AEC stations out to 420 fathoms, about as deep as the *Cobb* could go with the standard equipment she had. The AEC stations would be sampled four times a year based on the four seasons.

Drawing 12 shows the location of proposed AEC stations made during Cruise 50 by their haul numbers, 60 to 74. The vessel went farther north, looking for another area to extend the depth of the stations, and made haul 74, which was the same depth as haul 73. The depth contours were to extend out to 1,000 fathoms and would be established later on. All the AEC tows were made for an hour, which made it easier to compare them. All the hauls were successful with the exception of haul 67 (AEC station 275), which snagged. Haul 74, the 420-fathom station, was successful but had only a few idiots and sablefish in the catch and was way north of the established stations.

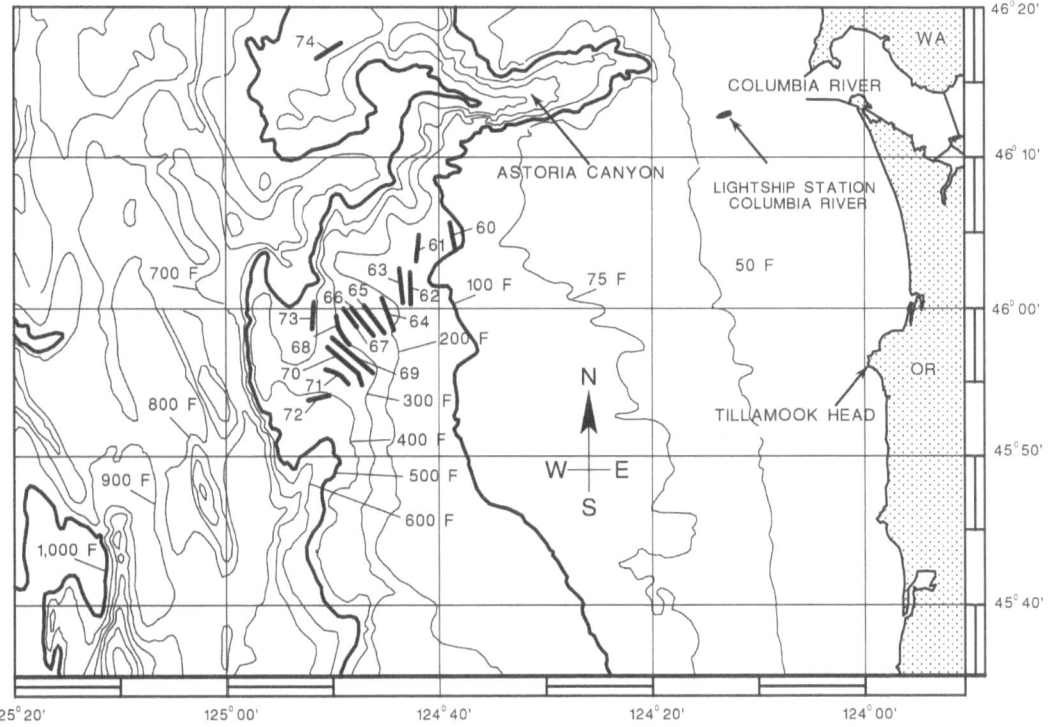

Drawing 12 Cruise 50—Start of AEC Track Line.

This was when I started to understand which species were accessible to the trawl on the continental slope by depth interval from 100 to 420 fathoms. Table 4 lists the major species taken at each AEC station. Dover sole were found in all fourteen stations. They were also found in commercial quantiles on the shelf during Cruise 46, a trip to explore an area on the shelf that was considered untrawable. Sablefish and the idiots were also found in almost all the hauls along the continental slope out to 425 fathoms, the deepest haul (73). Haul 67 was unsuccessful because it hung up and part of the catch was lost, so it was not included in this table. It is fascinating to see how deep the rockfish reached, 250 fathoms, with POP reaching 225 fathoms. Lee and Al had established the start of the AEC track line to at least 425 fathoms.

Table 4 Major Species Taken During AEC Track Line Set Up—Cruise 50.

Haul No. Cruise 50	60	61	62	63	64	65	66	68	69	70	71	72	73	74
AEC Station	3	4	5	6	7	8	9	11	12	13	14	15	16	16
Depth Fathoms	100	125	150	175	200	225	250	300	325	350	375	400	425	425
English Sole		X												
Dover Sole	X	X	X	X	X	X	X	X	X	X	X	X	X	
Rex Sole	X	X			X	X								
Turbot	X	X	X		X	X	X							
Sablefish	X	X	X		X	X	X	X	X	X	X	X	X	X
Brown Shark				X	X	X	X	X		X			X	X
Dogfish	X	X	X		X	X	X							
Idiot		X	X	X	X	X	X	X	X	X	X	X	X	X
Black Rockfish														
S. brevispinis	X	X												
S. flavidus	X													
Red Rockfish														
S. aleutianus						X	X							
S. alutus, POP	X	X	X	X	X	X								
S. crameri		X	X	X	X									
S. diploproa		X		X										
S. elongates	X													
S. helvomaculatus		X												
S. paucispinis		X	X											
S. saxicola			X	X										

Appendix 9 lists the thirteen successful stations made along the AEC track line. Haul 67 was excluded from the list because it hung up, and haul 74 was also excluded since it was a duplicate of haul 73 at the 425-fathom station and way north of the track line. The appendix also lists the catch of rockfish taken in the thirteen hauls made on the slope. The haul numbers are given, as well as the AEC station. Two species of black rockfish, *S. brevispinis* and *S. flavidus*, were found in the shallower waters of the slope in trace amounts. There were eight species of red rockfish. POP *(S. alutus)* was found in six hauls from the top of the continental slope at 100 fathoms to the 225-fathom station. There were four hauls from 125 to 200 fathoms. The catches were large, ranging from 1,200 to 3,500 lb. of POP. *S. aleutianus* catches were 150 to 400 lb. in two hauls, 65 and 66, in waters of 225 and 250 fathoms, the deepest water that caught any rockfish. The rest of the rockfish catches were small. The idiots were found in all the catches from 125 to 425 fathoms, the deepest.

Adding the hauls made for the AEC track line to those made in the first half of the cruise brought the total successful hauls for Cruise 50 to thirty-three and the individual rockfish species to thirteen. They are all listed in Table 5. POP dominated the catch occurring in twenty-one of the total hauls, followed by *S. crameri, S. rubrivinctus,* and *S. diploproa*. *S. rosaceus* was not seen during the 1961 survey.

The vessel returned to Seattle, completing Cruise 50. I had a better understanding of how fish lived along the bottom of the continental slope. I was happy to be back in the office once again after completing my part of the two cruises I was on. The time spent on them was exciting and stimulating, but the time spent between was busy. There didn't seem to be enough hours to complete the assigned tasks before departing on the next cruise.

Table 5 Rockfish Species Taken During Cruises 13 and 50.

Year	1961	1961	1961	1961		1952
Cruise No.	CR # 50	CR # 50	CR # 50	CR # 50		CR # 13
Appendix	7-9	7 & 8	7 & 8	9		
Depth Fathoms	78.5-287	78.5-98.5	100-287	100-425		100-225
No. of Hauls	*33	4	16	13		32
Continental	Both	Shelf	Slope	Slope		Slope
Black Rockfish						
S. brevispinis	**9	3	4	2		–
S. entomelas	3	1	2	–		–
S. flavidus	4	2	1	1		–
Red Rockfish						
S. aleutianus	2	–	–	2		X
S. alutus, POP	21	1	14	6		X
S. crameri	13	–	9	4		X
S. diploproa	10	–	8	2		X
S. elongatus	6	3	2	1		–
S. helvomaculatus	1	–	–	1		–
S. paucispinis	8	2	4	2		–
S. pinniger	5	3	2	–		–
S. rosaceus	–	–	–	–		X
S. rubrivinctus	12	1	11	–		–
S. saxicola	7	–	4	3		X
No. of Species	13	8	11	10		6

*The thirty-three successful trawl hauls made during the cruise.

**The total number of hauls each species were found in the thirty-three.

Rockfish X present — not present.

Chapter 6: Endnotes

1. C. R. Hitz and D. L. Alverson, "Bottom Fish Survey off the Oregon Coast, April-June 1961," *Commercial Fisheries Review* 25, No. 6 (June 1963). MS #89.

2. Milton S. Love, Mary Yoklavich, and Lyman Thorsteinson, *The Rockfishes of the Northeast Pacific* (Berkley: University of California Press Berkley, 2002), pp. 114–116. *Sebastolobus alascanus*.

3. A. T. Pruter, "Field Party Chief's Notes Cruises 50 Bottom Fish Survey off the Oregon Coast, April–June 1961," Records of the *John N. Cobb*, Historical Ship Files, ID 119654305, US National Archives Seattle.

CHAPTER SEVEN

Between Trips

Returning to the office after a trip was interesting. The ground kept moving while I was standing on solid ground. Apparently, the body still thought it was at sea, but after a few days, I adjusted and became stable again. While at the College of Fisheries, I heard rumors about individuals getting seasick when they set foot on dry land after getting their sea legs on a trip where they encountered severe weather. It was a busy time. The first item that occurred after a cruise was the debriefing, a meeting with Director Lee Alverson, the CS, and the skipper of the *Cobb*, Pete Larsen. They discussed any problems with the crew, the operation of the vessel, and how the scientific party fit in with the crew. If there was a problem, the director had the authority to solve it, which he did. The second item was to prepare the cruise report, which was issued and distributed to the public. It mainly went to the fishing industries in the states of Washington, Oregon, and Alaska and contained pertinent information about the cruise, including successful tows and catches for a bottomfish cruise. The final thing was to select the individuals who would prepare the final report to be published upon completion. A manuscript number was assigned. The office had a manuscript file (MS) to keep track of all the publications published by members in the office. Their MS file is stored at the US National Archives at Seattle under the Historical Ship Data Files—Records of the *John N. Cobb*.[1]

The decision was made to combine Cruises 46 and 47 into one manuscript, MS 60, and I was assigned the task. It was published in June 1961 in the *Commercial Fisheries Review* and was titled, "Bottom Trawling

Explorations off the Washington and British Columbia Coasts, May–August 1960." It gave me a lot more knowledge on POP and rockfish after leaving the College of Fisheries. After I returned from Cruise 50, I was assigned to complete its final report, MS 89. Both had the same objective of exploring the bottom that commercial fishermen considered untrawlable. It was published in June 1963 in the *Commercial Fisheries Review* and was titled, "Bottom Fish Survey off the Oregon Coast, April–June 1961." One of the items that was left out of the two MS's was how abundant was each species of rockfish to the commercial otter trawl.

Summary — Lee's Rockfish Cruises

Once Lee was hired the first time with the exploratory base in 1950, he conducted two deepwater surveys using the *Cobb*, starting with Cruise 9 and followed by Cruise 13. After he quit the base and was employed by the State of Washington for five years, he was again hired by the exploratory base as the director. He scheduled three rockfish cruises, 46, 47, and 50. After I was hired in 1960, I was involved in the three cruises.

There was a total of eighty-two successful hauls made in these three cruises. Each species that was found in the catch is listed in Table 6, along with the number of hauls they were found in. The top four most abundant were *S. brevispinis,* found in fifty-two of the hauls (63.4 percent), followed by POP (*S. alutus*) in forty-seven hauls (57.3 percent), *S. pinniger* in forty hauls (48.8 percent), and *S. paucispinis* in thirty-four (41.5 percent). There wasn't much information on what species occupied the slope; therefore, the eighty-two hauls were divided into shelf and slope, with the dividing line at 100 fathoms, known as the continental break (Table 6). Three species were found in the hauls made on the shelf side only: *S. melanops, S. proriger,* and *S. wilsoni*. There were five species found on the slope side only: *S. aleutianus, S. crameri, S. diploproa, S. helvomaculatus,* and *S. zacentrus*. The rest of the species were found on both sides of the line. Regarding the top four, *S. brevispinis* dominated the shelf side with forty hauls and had twelve hauls on the slope side, POP dominated the slope side with twenty-six hauls versus twenty-one on the shelf, *S pinniger* was in thirty-five hauls on the shelf and five on the slope, and *S. paucispinis* was in twenty-three hauls on the shelf and eleven on the slope.

Table 6 Rockfish Species Taken During Cruises 46, 47, and 50.

Year	1960	1960	1961	Total		
Cruise No.	CR 46	CR 47	CR 50		Shelf	Slope
Area	Spit	BC	OR			
Depth Fathoms	57-92	58-115	78-425		55-99	100-420
No. of Hauls	34	15	33	*82	47	35
Black Rockfish						
S. brevispinis	30	13	9	**52	40	12
S. entomelas	–	2	3	5	1	4
S. flavidus	16	7	4	27	21	6
S. melanops	3	–	–	3	3	–
Red Rockfish						
**S. aleutianus	–	–	2	2	–	2
S. alutus, POP	16	10	21	47	21	26
S. crameri	–	–	13	13	–	13
S. diploproa	–	–	10	10	–	10
S. elongatus	7	4	6	17	13	4
S. helvomaculatus	–	–	1	1	–	1
S. paucispinis	19	7	8	34	23	11
S. pinniger	27	8	5	40	35	5
S. proriger	4	2	–	6	6	–
S. ruberrimus	5	8	–	13	10	3
S. rubrivinctus	14	2	12	28	16	12
S. saxicola	–	5	7	12	1	11
S. wilsoni	1	–	–	1	1	–
S. zacentrus	–	3	–	3	–	3
No. of Species	11	12	13	18	13	15

*The eighty-two successful trawl hauls made during the three cruise.

**The total number of hauls each species were found in the eighty-two.

S. *brevispinis* was the first species of rockfish available to the trawl, being caught in fifty-two of the eighty-two hauls taken during the three surveys. They also dominated the catch on the shelf with forty hauls. There were seven large catches of 1,000 lb. or more. Hauls 16, 19, and 20 were made during Cruse 46, all on the shelf (Appendix 3). The other four were made during Cruise 47. Haul 40 in waters of 104.5 fathoms was made just over the break on the slope (Appendix 5), whereas the other three were taken on the shelf side, hauls 48, 50, and 51 (Appendix 6). The depth range where *S. brevispinis* was found in was from 58 fathoms (haul 39, see Appendix 5) to as deep as 132.5 fathoms (haul 20, see Appendix 8) on the slope. The total catch of *S. brevispinis* was 20,474 lb. out of a grand total for all species of 171,014 lb. or 12 percent of the total catch by weight.

POP was second regarding the total chatch with forty-seven of the eighty-two hauls made. There were twenty one hauls on the shelf side and twenty six on the slope side, making POP the dominant species on the slope. Regarding large catches of 1,000 lb. or more, there were nineteen hauls. Six of the hauls made during Cruise 46 were on the shelf close to the break—hauls 28, 29, 30, 33, 34, and 37—in water depths of 77 to 88 fathoms (Appendix 4). The next five hauls were taken during Cruise 47. The first three—hauls 40, 41, and 42—were made just before the break in 70 to 75 fathoms (Appendix 5). The other two hauls, haul 55 was made close to the break in 96 fathoms, and haul 53 was just over on the slope in 112.5 fathoms (Appendix 6). The remaining eight hauls were made on the slope during Cruise 50. The first one was a gigantic haul of 11,000 lb. that got the name the "Perch Spot," which was haul 20 in 132.5 fathoms. The other three were hauls 23, 25, and 28 made in water depths of 132.5 to 138.5 fathoms (Appendix 8). The last four large catches of POP were taken during the establishment of the AEC stations, part of Cruise 50. They were hauls 61 to 64 and were made in water depths of 125 to 200 fathoms (Appendix 9). There is no question that *S. alutus* is the dominant species, especially on the slope. Out of the total catch of 171,014 lb., the weight of POP was 59,669 lb. or 35 percent of the total catch by weight. POP were found in water depths of 71 to 183 fathoms. The best catches occurred between 75 and 138.5 fathoms.

S. pinniger was the third species found in forty of the eighty-two hauls. They were second with thirty-five hauls on the shelf side, with five hauls of 1,000 lb. or more made during the three cruises. Hauls 16 and 19 were made on the shelf in about 74 and 77 fathoms (Appendix 3). The last three were made near the break. Hauls 51 and 52 were made at 92 and 104 fathoms (Appendix 6), and haul 21 was made at 100 fathoms (Appendix 8). The total catch was 9,346 lb. or 5.5 percent of the total catch of 171,014 lb. There were eighteen species of rockfish taken during the three cruises.

S. brevispinis and *S. pinniger* were the rockfish that dominated the shelf while *S. alutus* dominated the slope.

Lee Alverson's Document

By 1963, Lee Alverson had a tremendous amount of information from the cruises I was on and all the other bottomfish information collected by the Exploratory Fishing Base from the Chukchi Sea to Southern Oregon. All the information was used in document MS 94, "A Study of Demersal Fishes and Fisheries of the Northeastern Pacific Ocean," which was published in 1964.[2] The study covered the depths of 0 to 600 fathoms and was subdivided into eight 50-fathom intervals. The species were subdivided into four groups: flatfish, rockfish, cods, and others.

Chapter 5 of this document covered the rockfish group. Rockfish represented the greatest number of species of any other single group, twenty-six different species of the genus *Sebastes*. There was a marked decline in the number of species occurring from south to north: twenty-two in Oregon and Washington, twenty in British Columbia and Southeast Alaska, thirteen in the Gulf of Alaska, and six off the Alaska Peninsula. The genera *Sebastolobus*, which the fishermen referred to as idiots, were included in the rockfish group and were found in all depth ranges out to 699 fathoms.

POP was the most important single species of rockfish harvested in the northeastern Pacific, dominating the rockfish catches on the outer continental shelf and upper continental slope in regions between Southern Oregon and Unimak Pass. It was the dominant species, contributing to the rockfish catches in regions south of Cape Spencer in depth zones from 50 to 299 fathoms.

Publish or Perish

I remember Lee Alverson saying, "If an item isn't published, the material is lost forever." That was where the statement "publish or perish" came from in our office. I felt that to get future raises, I would follow the unwritten office policy; "Go to sea and publish, and you will be rewarded" was always on my mind while I was with the exploratory fishing group.

On my first trip, I learned something that helped me out during those years. We spotted a Japanese fishing float in Queen Charlotte Sound, so we slowed down and ran alongside it. One of our deck crew took a large hoop with a net on a long pole, scooped the float out of the water, and brought it aboard. It had a net around it, with gooseneck barnacles attached to the webbing, and had been floating for some time because smaller barnacles were attached to the larger ones in an upside-down pyramid. With time, they would have sunk the ball. When it was brought aboard, eight small rockfish of two different species less than two inches in length fell out of the mass. After we got back, I wrote and published a note about them.[3] After that, I kept looking for items to publish while at sea.

I was assigned an office space for two in the Montlake building. On my side, there was a desk with a full-size drafting table next to it, something I cherished and used regularly since a drawing is worth one thousand words, which would help with the publishing part of the unwritten office policy. When I was an undergraduate in college, I took drafting classes as my elective and got an A in them, while all my other classes were Cs, with a few Bs. The professor tried to talk me into changing my major, but I had other plans.

The US tuna fleet was going through a radical update to its vessels as fishing gear and methods of capturing tuna had changed. They were converting the tuna clippers or the pole and line fisheries to purse seiners. Dick McNeely, part of the exploratory staff, had just returned from an assignment at the time I was hired. The changes were based on the introduction of the power block, which reduced crew size, and nylon nets, which would not rot in the tropical sun like cotton ones did. I was interested in the fleet conversion and did several drawings for McNeely's publication, "The Purse Seine Revolution in Tuna Fishing," published in the June 1961 issue of *Pacific Fisherman*.[4]

The books Lee brought back from the European meeting on world fishing were fascinating, and I spent time between trips reading them. One presentation was titled "Pacific Combination Fishing Vessels."[5] The *Cobb's* design on the outside was a typical Pacific coast combination vessel. The articles written by H. C. Hanson, a marine architect, were interesting because he described the design concept. He had also designed the four fishing vessels that accompanied the *Pacific Explorer*. I spent time looking at other articles and detailed drawings, sketching the profiles of side trawlers and stern trawlers of the world's fleets. In 1969, I was assigned to write an article in the *Encyclopedia of Marine Resources* entitled "Fishing Vessels and Support Ships."[6] After it was published, a review appearing in the March 1970 issue of *Science* stated, "These faults should not detract from the usefulness of many of the articles: the drawings of fishing vessels by Hitz are a pleasure to look at as well as being clear and informative."

Rockfish Field Key

There were several projects I wanted to work on when I was back at the office, when I had time between assigned jobs. One was the development of a field key for identifying individual species of rockfish at sea.[7] I selected the fifty-three species of rockfish I believed existed along the northeastern coast of the Pacific at the time, placing each name on a standard-sized typing page in portrait orientation. Dividing them into four categories by body color—striped rockfish, black rockfish, red rockfish, and white/red-spotted rockfish—I placed each under a tab representing the color of each group along the outside margin of the page, starting with the striped near the top and proceeding down the side and ending near the bottom with the spotted.

Colored photos would be ideal, but at the time, printing color photos was too expensive. I decided that a pen and ink scale drawing of the profile of each fish was the way. I used Phillips's publication as a guide to making the drawings, using the black-and-white photos he had of individual fish to get the outlines, measurements, and counts he listed for each species to ensure the accuracy of the drawing.[8] I placed the drawing of an individual rockfish in the middle of the page. Selecting one or two characteristics that separated an individual from the rest of the group, I condensed its description to a few words and placed it on the drawing next to the profile of the fish. A line was drawn from the text to the profile so that the characteristics identifying the fish could be easily recalled in the field. It was finally published in March 1965, with a total of fifty-three species of *Sebastes* listed, and it worked![9]

After NOAA was formed in 1970, the observer program expanded, and individuals were trained to go out on foreign vessels to monitor those catches, taking one of the booklets with them. It was modified in 1977 and again in 1981, adding four species for a total of fifty-seven species. Additional copies were made over the years. In October 1998, the authors Orr, Brown, and Baker developed a new guide, which improved the old circular by replacing many of the drawings with color profile photos of each species of rockfish. In the introduction, Orr states, "Primarily designed as an aid in field identification, this guide follows the basic format of Hitz's (1965) 'Field Identification of the Northeastern Pacific Rockfish

(*Sebastes*),' the first guide to successfully use color as a major characteristic to identify species of *Sebastes*." In August 2000, a second revision was made, and almost all the drawings were replaced with color photos, making identification of individual fish easier.[10] Those publications listed a total of sixty-six species of *Sebastes* in the northeastern Pacific Ocean, increasing the count by thirteen from the fifty-three that I used in 1965.

Drags of the *John N. Cobb*

I and Pete Larsen, the skipper of the *Cobb,* would meet in the pilot house of the *Cobb* when it was in port and lay out a chart on the chart table and we developed a notebook which listed all the drags the *Cobb* had made.[11] It was an in-house publication that was circulated to commercial trawl fishermen. The original idea came from the European Fish Charts for the North Seas trawl fleet that had drag information printed on the chart. Knowledge that each Pacific coast fishermen had collected over the years was too precious for them to pass on to competitors, but information that the *Cobb* collected during the exploratory bottomfish cruises was public and the information became popular. The US Loran-C lines, an important navigation device that was printed on the charts, was discontinued in February 2010, making the used of loran lines and depth as the coordinates of individual drags obsolete. Now commercial fisherman have the use of digitized information and can use their computers with digitized charts and global positioning systems to establish individual drag information.

The end of all our onshore breaks between trips always came faster than expected. Again, the time came when I had to go back to sea, setting aside the jobs I was involved with until I got back. Once we left the dock and were underway, I made up my bunk and could started looking forward to a new trip and forgetting the shore time and chores around the house—they would have to wait.

Chapter 7: Endnotes

1. Records of the *John N. Cobb*, Historical Ship Files, ID 119654305, US National Archives Seattle. The Exploratory and Gear Research Base manuscript file (MS) had a list of all the manuscripts published or drafted by them. It is located at the Seattle US National Archives office near the NOAA Sandpoint Facilities. The Archives have converted the MS numbering system into their own numbering system.

2. D. L. Alverson, A. T. Pruter, and L. L. Ronholt, *A Study of Demersal Fishes and Fisheries of the Northeastern Pacific Ocean* (Vancouver: University of British Columbia, 1964). MS #94.

3. C. R. Hitz, "Notes: Occurrence of Two Species of Juvenile Rockfish in Queen Charlotte Sound," *Journal of the Fisheries Research Board of Canada*, 18(2) (1961), pp. 272–281. MS #61.

4. Richard L. McNeely, "The Purse Seine Revolution in Tuna Fishing," *Pacific Fisherman* (June 1961), pp. 27–58. MS #63.

5. H. C. Hanson, "Pacific Combination Fishing Vessels," in *Fishing Boats of the World*, ed. Jan-Olof Truang (London: Arthur J. Heighway Publications LTD, 1955), pp. 187–199.

6. C. R. Hitz, "Fishing Vessels and Support Ships," in *Encyclopedia of Marine Resources*, ed. Frank E. Firth (New York: Van Nostrand Reinhold Co., 1969), pp. 250–260.

7. Charles R. Hitz, "Field Rockfish Key," Carmel Finley (blog), November 9, 2016. https://carmelfinley.wordpress.com/2016/11/09/field-rockfish-key. Bob's posting #41.

8. J. B. Phillips, "A Review of the Rockfishes of California (Family Scorpaenidae)," *California Fish and Game Fish Bulletin* No. 104 (1957).

9. C. R. Hitz, "Field Identification of the Northeastern Pacific Rockfish (*Sebastodes*)," *US Fish and Wildlife Circular* 203 (1965).

10. J. W. Orr, M. A. Brown, and D. C. Baker, "Guide to Rockfishes (Scorpaenidae) of the Genera *Sebastes*, *Sebastolobus*, and *Adenosebastes* of the Northeast Pacific Ocean," NOAA Technical Memorandum NMFS-AFSC-96 (1998).

11. Charles R. Hitz, "The Perch Spot and the Fisherman's Black Book," Carmel Finley (blog), October 31, 2021. https://carmelfinley.wordpress.com/2012/10/31/the-perch-spot-and-the-fishermans-black-book/. Bob's posting #4.

CHAPTER EIGHT

Gulf of Alaska Cruises 1961–1962

The objectives during the next four trips I was assigned differed from my first two bottomfish exploratory trips, Cruises 47 and 50, where the objective was to find successful trawl hauls in areas commercial fishermen considered untrawlable. There were two new objectives. For Cruises 52 and 54, we were to survey the Gulf of Alaska to see what fish were taken in a commercial otter trawl. During Cruises 53 and 57, we were to monitor the set of AEC stations off the Columbia River during the winter. Not only did the objectives change, but they overlapped one another during the years 1961 and 1963. Therefore, they are discussed in two parts: chapter 8—Gulf of Alaska Cruises, and chapter 9—Columbia River AEC Cruises.

The survey of the Gulf of Alaska was part of a joint study by three different units: the Seattle Exploratory Fishing Base, the newly established Juneau Exploratory Fishing Base in Alaska, and the International Pacific Halibut Commission (IPHC). I was assigned to two of these surveys: the first half of Cruise 52 in 1961 and the last half of Cruise 54 in 1962. At the time, I wondered why our methods had changed from looking for successful hauls in a defined area classified as untrawlable by commercial fishermen to making a successful haul in each square of the study that divided the Gulf of Alaska. We used the sounder to determine if the bottom was trawlable in the square. If it was, we would try the net for an hour tow. If the trawl was successful, we would record what fish were taken.

I was assigned MS 92 "The Bottom Trawling Surveys of the Northeastern Gulf of Alaska," which covered Cruises 52 and 54.[1] I was focused on the manuscript and only mentioned that we were in cooperation with the IPHC. It wasn't until I read Carmel Finley's April 19, 2017, blog "The Further Scientific Career of the *Western Flyer*" that I finally realized why we were involved with IPHC, and it wasn't just an exploratory cruise.[2] Colin Levins was a biologist aide that was hirded by the IPHC. He was on board the *Western Flyer* when it was chartered by the Pacific Halibut Commission in 1962-63. The article which he wrote for the *Argeonauta,* which Carmel gave use a link to, is the story of his experiences on the vessel.[3] He gave the objective of what the IPHC was after: what effect trawling would have on the sustainability of the halibut in the Gulf of Alaska. The Soviet fleet was beginning to trawl in the Gulf of Alaska along with Japanese trawlers. The halibut resource had been well-managed through the years when the commission had restricted trawlers from marketing halibut. They could only be taken by longlines. The Gulf of Alaska was a primary area where Canadian and US halibut long liners caught halibut, but what results would trawl fisheries have on the halibut resource? It was a much bigger affair than I realized, as after all these years, I thought we were still exploring the Gulf of Alaska for possible new resources and that the Halibut Commission was just adding to our study. Focused on the *Cobb's* trips and not the big picture, I was surely naive to think that the only two vessels involved in the surveys were the Seattle exploratory vessel *John N. Cobb* and Juneau Alaska's Exploratory Office's chartered vessel *Tordenskjold*. The Halibut Commission chartered other trawlers. For the 1961 survey, they chartered the *Arthur H., Morning Star,* and *St. Michael* and, for the 1962 survey, the *Arthur H., St. Michael,* and the *Western Flyer*.

 I assume there was a considerable amount of planning going into the study, but I was not involved. Al had worked with the Halibut Commission and must have had a lot to do with the organization, which differed from what we had done along the Washington, Oregon, and BC coasts. The sampling plan followed one that was worked out with the IPHC for exploring the entire massive continental shelf of the Gulf of Alaska region. The stations were located six miles apart along each line of longitude, which are spaced fifteen minutes apart. The lines were started in the west, and the stations on each were staggered as they moved east. Each square was designated as a station block, with the center defined as a station. The blocks were designated alphabetically, the closest inshore as A, moving outward along the line to a depth of 250 fathoms. Lines 59 to 82 were assigned to the *John N. Cobb* (Drawing 13), and lines 83 to 114 were assigned to the charter vessel *Tordenskjold*.

Each square was to be sampled with an otter trawl somewhere within the square. The vessel would sound the square to determine if the bottom was trawlable. If it appeared level and soft, a one-hour haul was attempted. Many of the blocks were determined untrawlable, while others resulted in snags. The catch was brought aboard from those that were successful and sorted by species in the normal procedure we had used off the Washington and Oregon coasts.

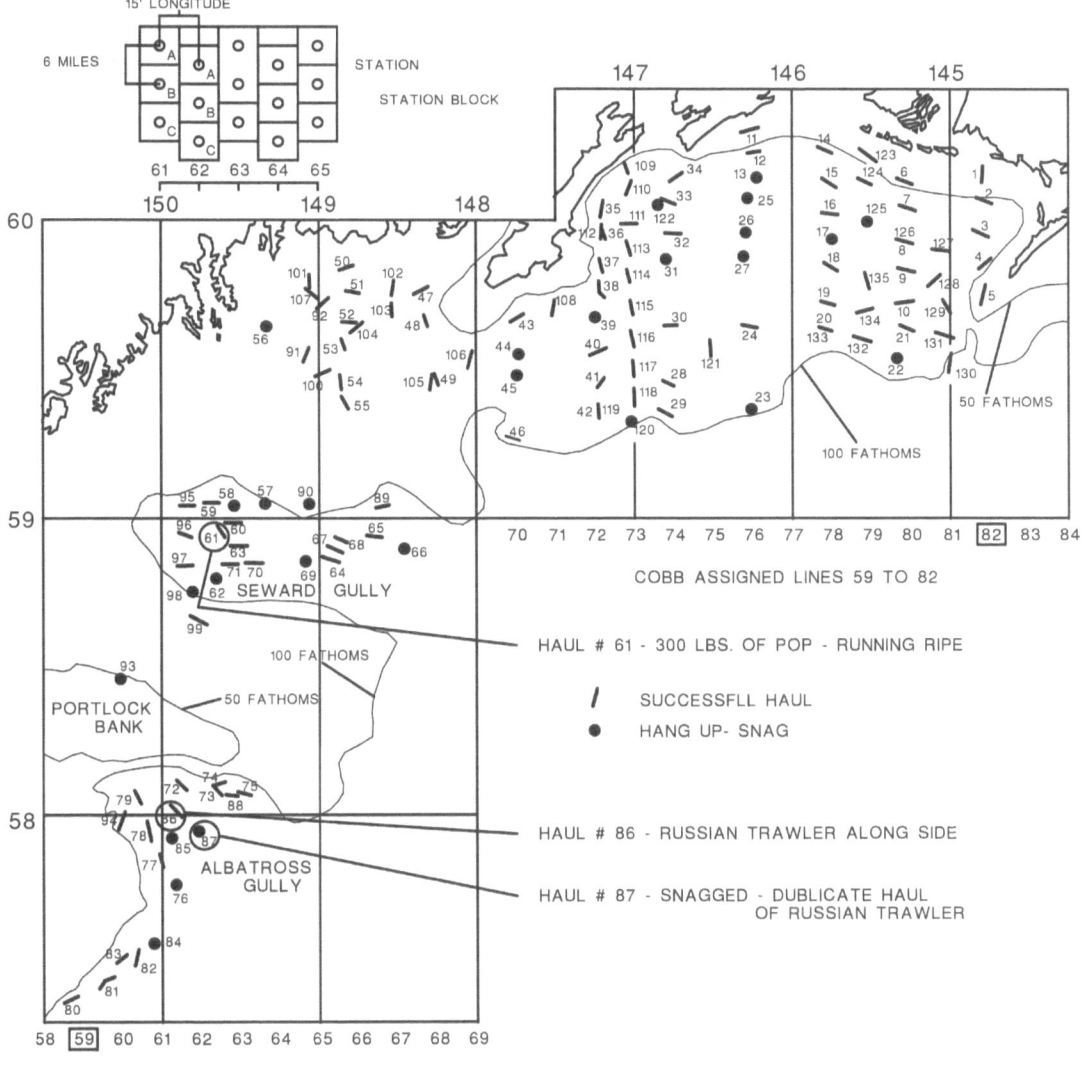

Drawing 13 R/V *John N. Cobb*—Cruises 52 and 54: Alaska Survey.

Since this study was conducted over known halibut grounds, all halibut weighing less than 40 lb. were placed in a live tank from which the viable ones were removed, tagged, and released. Larger halibut not placed into the live box were measured, tagged in the checker, and returned to the sea immediately. Untagged halibut were measured, had their sex determined, and had their otoliths removed for age and growth studies in the future by IPHC biologists.

The halibut is an interesting flatfish that lies on its side on the bottom of the ocean. The eye on the left-hand side migrates to the right side in the larval stage of growth as it adapts to the ocean bottom. Once the two eyes become permanent on the right side, dark pigment forms on that side, becoming camouflage when the fish is lying on the bottom. The other side is white.

They are a large fish. In fact, they are the largest boney fish in the world, reaching a total length of 9 feet and weighing close to 500 lb. The average commercial landing size is about 35 lb., and they have been sought after by Canadian and American fishermen since 1892. In 1924, a convention formed between the two countries resulted in the best-managed fishery in the world. One requirement was that only longlined or troll-caught fish could be marked, outlawing trawl-caught fish.

Cruise 52: My Third Trip (1961), First Half

The *John N. Cobb* left Lake Union on September 5, 1961, with Al Pruter, the CS, and myself on a new project. It was also a new adventure for me, a trip to Alaska! After taking on fuel and heading through the locks, we entered the strait through Admiralty Inlet. I had made up my bunk, and we cleared the inlet. The tablecloth remained on the galley table, and the railings for the table remained in storage since we did not turn west out of the strait but kept going northwest, headed for the protected waters of the Inside Passage. I didn't worry about seasickness until we reached Cape Spencer, the end of the Inside Passage, and entered that huge expanse of water of the Gulf of Alaska.

We passed Cattle Point on San Juan Island, where my brother and I used to hunt rabbits, which had multiplied after an individual on the island let his domestic rabbits go. The farmers wanted them removed and encouraged hunters to hunt them on their property. During the Second World War, the farmers would capture them at night using small trucks with spotlights. The rabbits froze in the beam of light and were netted and caged. They would be sold on the mainland, providing yet another food source for the hungry nation at war.

I stood on the bow and watched the constant swirl of the changing tidal waters between the two massive parts of the Salish Sea, the Strait of Juan de Fuca and the Strait of Georgia, as we passed from one to the other, traveling between the Canadian and American San Juan Islands. In the Strait of Georgia, I respected these waters. There were many stories of fishing boats perishing while transiting to and from Alaska. We could see what could happen to vessels traveling the straits when a blow came up from the southeasterly direction—the seas could get high, especially on an outgoing tide, which is probably where the stories of lost vessels came from. We reached the end of the straits the next morning and passed through the Seymour Narrows at slack water at 8:30 a.m. This is a narrowing where the tidal currents can reach fifteen knots. Every six hours, there is a change of tide, or slack water, when a vessel can pass through with no problems. But even at slack water, there is never a time of quiet; the water is constantly moving. In 1958, the top of Ripple Rock in the Seymour Narrows was blasted off, reducing the danger to ships passing through.

Once through the narrows, it seemed we were in a channel that went on forever. When we came to a Y, the skipper would choose one or the other. I asked him if he ever made a wrong choice and the channel came to an end, but he had never done so. Human populations had disappeared along with buildings, and trees now covered the sides of the mountains from the water to the snowcapped peaks, with the *Cobb* pressing north between them. We finally broke out into Queen Charlotte Strait, passing Port Hardy, where we had been on my first trip.

Once again, we were in open water, where we could feel the ocean swell of the North Pacific. The *Cobb* ducked in behind islands, following the passage fish boats had used for years. It narrowed down to less than a quarter of a mile wide, with steep banks covered by trees and a breathtaking view that went on for miles until finally entering Dixon Entrance and passing out of Canadian waters into Alaskan waters. We were halfway to the Gulf of Alaska and still had to pass through the inside waters of Southeast Alaska. I was surprised at the number of fish boats we met on our way north. The salmon fishing was closing down for the season, and salmon seiners, packers, and trap tenders were heading back south to Seattle. The most interesting vessels were the Canadian halibut long liners heading south for Vancouver, BC, loaded with halibut taken in the Gulf of Alaska. They were beautiful boats — modern combination vessels.

We finally reached Cape Spencer and the entrance into the Gulf of Alaska at about 4:00 p.m., away from the protected inside waters we had been traveling. We could feel the ground swell of the Pacific Ocean. Our course was west for Prince William Sound. We had traveled about four hours when we got a radio call from the *Tordenskjold*. The skipper asked Pete, our skipper, to plot a course for them to Cape Spencer from the loran readings they had just made. Pete plotted the position on the chart and calculated the course they should steer for Cape Spencer. The *Tordenskjold* had been on charter, and when that ended in Kodiak, the CS had left the boat with the chart. It had all the stations plotted on it, and unfortunately, there was not another copy on board. Once they got to Cape Spencer, they would be fine.

The good weather deteriorated as we crossed the gulf. The wind picked up all the next day, blowing thirty-five to forty knots by late evening. We anchored behind Montague Island at 2:00 a.m., at which time the wind was blowing forty-five knots with stronger gusts. Once the weather moderated, we proceeded to our first line of the sampling grid, line 82 (Drawing 13), starting at the shallowest station, A. It was sounded and was found trawlable. The first haul was made successfully and was plotted on the chart. The following hauls, 2 to 5, were made in each of the following squares, completing the first five stations, A to E. The last haul, 5, was just southwest of Cape St. Elias, but catches for all five were poor.

Catches were poor throughout most of our half of the cruise until we got into deeper water. Haul 40, on line 72 at 115 to 118 fathoms of water, was the deepest haul so far, and the catch of POP was 320 lb., an indication of their presence in the deeper water. On the last haul of our half of the cruise, haul 42 on line 72, we took 1,167 lb. of POP. I was disappointed in the rockfish catch when the hauls we were involved in were less than 20 lb. We tagged 156 halibut by the end of our half of the trip. I remember on one of the successful tows, we came up with several large halibut. It was my first experience with these large and beautiful fish. I was standing in the checker, straddling the fish, getting ready to tag one by placing the tag on the upper side of the gill cover on the eyed, or dark, side. I felt a tap on my leg. It was Cony Mohan, one of the deck crew, a Norwegian with years of experience as a halibut fisherman before he came to the *Cobb*. He was small and wiry and had the most wonderful blue eyes that could see things on the horizon I couldn't distinguish. He said, "Bob, if you want to keep your manhood, do not straddle that halibut when you put the tag on." Then he showed me that if you turn the halibut over with the white side up, it lies there as quietly as if asleep. If you rubbed its belly, you would swear you could hear it purr like a cat. I then measured the halibut for total length, rolled it over, and placed the tag on the gill cover from outside the checker. Suddenly, it came to life. If I had straddled it to put on the tag, I would surely have lost my manhood and then some. I am glad to have taken Cony's advice.

Our part of the cruise finished, we had completed six lines: 72, 74, 76, 78, 80, and 82. The lines were staggered. It was a two-year project, and the missing lines would be surveyed the next year during Cruise 54. We found forty-two of the squares possible to trawl. Thirty-three were successful, and nine snagged or hung up. Al and I flew out of Cordova on October 3, 1961, on a Douglas DC-4, a four-engine propeller-driven transport to Seattle, before the jets arrived in Alaska in the mid-1960s. That was my first flight on a large commercial aircraft. It was a thrill for me as it was much faster than the *Cobb*. I was finally home again.

The second half of the cruise, which I wasn't on, completed lines 64, 66, 68, and 70. They sounded each square, finding fourteen possible tows. Eleven were successful, and three hung up. The catches were about the same as the first half, generally poor.

Cruise 54: First Half (1962)

The CS for the first half of the trip was Al Pruter. I wasn't aboard, but I did review his field notes, which helped me understand what happened regarding the POP and the Russian trawlers during his part of the trip. It started when the *Cobb* arrived in Juneau on April 20, 1962, after transiting the Inside Passage from Seattle. Al boarded the vessel, which proceeded across the Gulf of Alaska, starting on line 64, which ran through the Seward Gully (Drawing 13). The line had already been sounded during Cruise 52 the year before. There were two possible hauls that were not tried because of bad weather. Al tried hauls 56 and 57 when he arrived, but both hung up. He then moved west to line 63, where he found several successful hauls. One of them, haul 61, made in water depths of 131 fathoms (786 feet), fascinated me when I read his personal log.[4] That haul had what I had been looking for ever since I left college. I was interested in determining the spawning time of the rockfish and had been examining ovaries of female rockfish that were taken on any cruise I was on. I was looking for the change in the color of the ovaries in the females once the eggs were fertilized. Thousands of embryos developing in the ovary change color from yellow to orange as the embryo grows, then to black when the eyes appear and the egg or yolk becomes smaller as it's used up. Once the yolk is gone, the mass of larvae are released into the sea to fend for themselves.

I found in his log dated for the day April 23, 1962, 100 lb. of large POP caught in haul 61. This was what I was looking for, the black stream looked like graphite running out of the female's vent, thousands and thousands of embryos spilling out on the deck from a single fish, their tiny eyes sparkling in the dark gray-black mass. An amazing sight! I saw it when I was doing my graduate work on the *Commando* and had never seen it again. Later, the *Cobb* moved from Seward Gully to the south to Albatross Gully. On line 61, haul 76, made at a depth of 125 fathoms, hung up, but it had the largest catch that had been seen so far—3000 lb. of POP. Then on May 5, 1962, in the late afternoon, after making haul 86, the *Cobb* observed a Russian trawler make a tow. It was an hour and a half long, with an estimated catch of 15,000 lb. It appeared to be a pure catch of POP and was dumped through the hatch to the fish hole. The *Cobb* had followed behind and, with the sounder, observed the net fishing along the bottom. The *Cobb* had shown hard bottom, which we could not trawl successfully. After the tow, the Russian passed closely by the *Cobb,* coming alongside close enough for the crews to exchange cigarettes and magazines. The Russians were very friendly. The Russian trawler moved off and met another Russian trawler, where they drifted to gather that night in the distance. The *Cobb* also drifted that night.

They were the first Soviet trawlers we encountered during the Gulf of Alaska surveys. There were two side trawlers about 140 feet long and had oval-shaped trawl doors (Fig. 18). The net was light construction, with eighty to one hundred steel floats on the head rope. There were about twelve weights spaced evenly along the foot rope, each on a 1-fathom-long line with an estimated 40 lb. sash weight attached to the end. The apparent purpose was to keep the net about 1 fathom off the bottom.[5]

The next morning, the *Cobb* observed one of the Russian ships trawling. The *Cobb* made haul 87, a duplicate of the two the Russians made that morning; however, the *Cobbs'* tow only lasted forty-four minutes, for it snagged and had to be hauled in early. The bottom depth was 118 to

Fig. 18 Soviet Vessel—Albatross Gully. Image #4590.

124 fathoms. The total catch was 812 lb., which included 550 lb. of POP. The Russian haul was observed when they picked up their gear, with an estimated catch of 3,000 lb. of pure POP. Apparently, they were flying their net just off the bottom. Al attempted to fly the *Cobb's* trawl by adding floats to the head rope. He made two tows on the same site as haul 87 and came up with 25 lb. of POP, but then he made a third tow with fewer floats that hung up, with a catch of 550 lb. of POP. With about four more days left of the first half of the trip, they looked for drags in Seward Gully and Albatross Gully. The *Cobb* then headed for Kodiak, took on fuel from the navy, and moved to the government pier, ending the first half of Cruise 54.

During the first half of the cruise, the *Cobb* surveyed four new lines 60, 61, 63, and 65. They completed partial surveys of lines 62, 64, and 66. They found thirty-seven possible hauls 57 to 94 of which twenty-seven were successful and eleven hung up or snagged. The catches were about the same, with the exception of hauls made in Albatross Gully, where the Russians were fishing.

Cruise 54: My Fifth Trip (1962), Second Trip as Chief Scientist

I was again assigned another Gulf of Alaska cruise for the second half, my fifth exploratory trip, and I was assigned as CS. Cruise 53—the winter AEC trip—was my first CS assignment. Apparently, I handled it well enough to be assigned to my second trip as CS. I was flattered they would even consider me in that role. On May 12, 1962, I flew into Kodiak, and the flight went well. It was my first flight on a Lockheed Constellation, another four-engine aircraft, which was similar to the Boeing B-17, one of my favorite WWII bombers.

I arrived at 9:30 a.m., met with Al Pruter at the Kodiak airport, and received instructions for the rest of the trip. He departed on the return flight to Seattle at 11:30 a.m. I then went to the *Cobb*, which was tied up at the government pier near the airport. For the next three days, the weather was extreme. We had moved from Kodiak and kept trying to get out to line 62 on Seward Gully and start working but had no success. On the fourth day, we finally made it out with a northwest wind at about twenty knots and a heavy southerly swell. The weather was too bad to do any fishing, but we were able to sound out possible tows. By evening, the sky was clear, the wind was about thirty knots, and there were heavy seas. We decided to sound all night so we could finish up the area by noon the next day. At midnight, the wind was at least twenty knots, seas were still high, and glass still coming up, so we waited till morning. Al had concentrated his effort on Seward Gully and Albatross Gully with Portlock Bank in between them, and he wanted us to see if we could get any more tows in the blocks in that area. The sounding showed the bottom to be hard but flat. Just before he had to go into Kodiak, Al tried one trawl haul in the area on line 60 (haul 93), and it hung up after two minutes on the bottom.

On May 20, the weather and waves finally came down, and we were underway to the grounds, setting haul 94 on line 60 on Albatross Gully. We moved to Seward Gully and made hauls 95 to 99 along line 62. Catches that day were primarily turbot. We drifted the rest of the night. Starting at 6:00 a.m., we sounded the west end of Seward Gully, but the bottom was no good for trawling. This finished up the area. We then went to lines 65 to 69 and picked off the squares that were trawlable—hauls 100 to 106 that were missed in the first survey. We were fortunate to have a good streak of weather and could get tows in the squares we missed the year before. We completed sounding six new lines—69, 71, 73, 75, 79, and 81—and picked up stations that were missed on lines 62, 65, 67, 68, 72, 74, and 80. We found forty possible hauls in the squares, thirty-six were successful and four were not. The catches were mainly turbot and pollock.

Weather in Alaska is always a concern, and I remember one day during this cruise where I had the best tea I have ever had served to me. We were out toward the end of the line with the gear out on a tow. About twenty minutes into it, the wind suddenly increased to a gale, the chop became waves, and the skipper ordered the trawl to be retrieved. By the time we got all the gear in and secured, the wind was howling, spray was everywhere, and the waves became much larger and started to break. We went inside and got out of our rain gear. I was wedged in the chair at the desk in the scientific stateroom, trying to relax and unwind. The cook poked his head around the door, said, "You need this," and handed me a cup of hot tea, then disappeared before I could thank him. My legs holding me in place, I held the cup in my hands, felt the warmth of the cup, and then took a sip. I must admit, it was the best tea I have ever had or will ever have. It was hot, and I was cold. I could taste the strong tea as well as the bourbon it was laced with, which warmed my insides. Government ships are free of alcohol, so I assume the cook felt I needed it, which I apparently did. Was it a medical emergency? I doubt it. After I relaxed and warmed up, I went up to the bridge. Reaching the top of the stairs, as the vessel's stern fell and the bow lifted, I looked out the port window to see monstrous waves, tops higher than the house. Thank goodness we were running with it.

We ended the fieldwork on the cruise with haul 135 on line 79. We put into Cordova, picked up the mail, then proceeded across the gulf with good weather and a following sea to Cape Spencer. Once we passed the cape, we were in the Inside Passage to Seattle. We made two stops on the way. At Juneau, we left samples we collected for the Auke Bay Lab. They brought the mail to the *Cobb,* and we purchased mess supplies from the local market. Our second stop was at Ketchikan, where we left a net off for the *Western Flyer,* who was charted by the Halibut Commission for the second year of the survey. The fishing gear was standardized between all the vessels used during the survey. We arrived in Seattle on June 8, 1962, ending the trip.

Lee and Al used this data along with the halibut charters for the project they were working on and the POP data they needed. Warren Rathjen, the director of Juneau's exploratory base, and I were assigned MS 92, which we published in 1965. It gave the results of the Gulf of Alaska cruise that the two exploratory bases were involved in. The *Cobb* covered lines 59 to 83, and the *Tordenskjold* lines 83 to 114. The rockfish catch from the two vessels comprised 11 percent of the total catch in pounds. It was divided into four groups: POP (*S. alutus*), idiots (*Sebastolobus*), rougheye rockfish (*S. aleutianus*), and other rockfish. POP made up 73.5 percent of the rockfish group and were found primarily in 101 to 150 fathoms of water. The idiots were found in all depths from 51 to 451 fathoms and made up 15.5 percent of the rockfish group. The rougheye rockfish made up 9.5 percent of the group and were found in deep waters of 201 to 250 fathoms. The other species of rockfish made up only 1.1 percent of the group and comprised seven species, which were divided into black rockfish *(S. brevispinis, S. flavidus, and S. melanops)* and red rockfish *(S. saxicola, S. crameri, S. helvomaculatus, and S. rubrivinctus)*.

Chapter 8: Endnotes

1. Charles R. Hitz and Warren F. Rathjen, "Bottom Trawling Surveys of the Northeastern Gulf of Alaska (Summer and Fall of 1961 and Spring of 1962)," *Commercial Fisheries Review* 27, no. 9 (1965). MS #92. See also Sep. No. 741.

2. Carmel Finley, "The Further Scientific Career of the *Western Flyer*," *Carmel Finley* (blog), April 19, 2017, https://carmelfinley.wordpress.com/2017/04/19/the-further-scientific-career-of-the-pacific-flyer/.

3. Colin Levings, "Chiefly between Kodiak Island and Cape Spencer, Alaska – a Memoir of Life on the Motor Vessel Western Flyer 1962-1963 and Influences on a Career in Marine Science," Argonauta 33, no. 3 (August 2016), pp. 5–19. http://www.cnrs-scrn.org/argonauta/pdf/argo_33_3.pdf#page=9.

4. A. T. Pruter, "Field Party Chief's Notes Cruises 54 Gulf of Alaska," Records of the *John N. Cobb*, Historical Ship Files, ID 119654305, US National Archives Seattle.

5. A. T. Pruter, "Soviet Trawler Observed in Gulf of Alaska," *Commercial Fisheries Review* 24, no. 9 (1962), pp. 11. MS. # 87.

CHAPTER NINE

Columbia River (AEC) Cruises 1962–1963

Early in 1961, the US Fish and Wildlife Service and the US Atomic Energy Commission (AEC) entered into an agreement to study bottomfish off the mouth of the Columbia River to a depth of 1,000 fathoms (6,000 feet). They wanted to know if any of the released material from the Hanford Site on the Columbia River would affect the commercial fisheries.[1] Track lines would be established from the lightship stationed off the Columbia River. Two possible series would be set up—an A series south of the Astoria Canyon and a B series north of the canyon. Bottom trawl stations would be located at 25-fathom depth intervals, starting at 50 fathoms. Once the track line was established, the drags were determined and set up as a permanent station. They would be sampled repetitively four times a year—spring, summer, fall, and winter—for two to three years.

At the end of Cruise 50, Lee Alverson and Al Pruter took the last few days of the cruise to lay out a series of stations off the Columbia River, which was the start of this project, and was considered the first, AEC-1 Cruise. The *Cobb* started on the south side, the A series, at 100 fathoms and proceeded out per 25-fathom intervals to 425 fathoms. Only the 275-fathom station was unsuccessful. It hung up. The university's R/V *Commando* was chartered from the UW College of Fisheries to conduct the spring AEC-2, summer AEC-3, and fall AEC-4 cruises, and the R/V *John N. Cobb* would conduct the winter Cruise 53, AEC-5. She was larger than the *Commando* and more suited for winter weather. I was assigned to the first AEC winter cruise. By the time I went out on the *Cobb*, the *Commando* had gathered samples at the established station and experimented with sampling the deeper stations out to 1,000 fathoms.

The techniques and equipment used during the study is outlined in *The Columbia River Estuary and Adjacent Ocean Waters*, published in 1972.[2] Part of chapter 18, pages 395 to 407, has the part I was involved in. Both the *Cobb* and *Commando* were upgraded for this work (Drawing 14). A hydro winch was put on each with enough cable to reach the bottom at the deepest station of 1,000 fathoms. A near-bottom water sample, along with a bottom temperature, was required to be made after each haul. A cannon ball was attached to the end of the cable and let out so the ball was just below the ocean surface. A person stood on the platform and attached a Nansen bottle with a reversing thermometer to the cable. Both ends of the tube are in the open position. It was then lowered to near the ocean bottom. A weighted messenger was placed on the cable and released, sliding down the wire and hitting a trigger on the Nansen bottle. The trigger closes both ends of a tube, closing and trapping the sea water sample taken near the bottom and freezing the thermometer reading. The hydro winch retrieved the Nansen bottle, the sample was preserved, and the temperature recorded.

The *Commando* received a new trawl winch similar to the *Cobb's*. The flat trawl doors were replaced with a new set of V-doors on both vessels. They were very stable and should reduce the occurrence of crossed doors. A new net reel was installed on the stern of both vessels, which was a great improvement for trawling (Drawing 14). The trawl bridles were rolled up on the net reel instead of storing them on the trawl winch, increasing the usable trawl cable. The net was attached to the bridles and rolled up on the reel also. The trawl doors were attached to the main trawl cables. The cod end of the net was placed over the stern, and the vessel slowly moved forward as the net and bridles rolled off the reel. Once the end of the bridle appeared, the reel was stopped. The net became a sea anchor, placing tension onto the flat link attached to the pennant attached to the net reel. The slack G-hook laying on the deck was attached to the after-door bridle by picking it up and attaching it to the flat link that was under tension. The net reel was then slacked off as the tension was transferred to the trawl doors. The slack G-hook and pennant was disconnected from the flat link and dropped onto the deck. The trawl doors were then let out until they were in the water, the winches were stopped, and the doors spread as they should. Then the cables were let out evenly to the proper depth. Once the haul was completed, the net was hauled in and the procedure was just reversed.

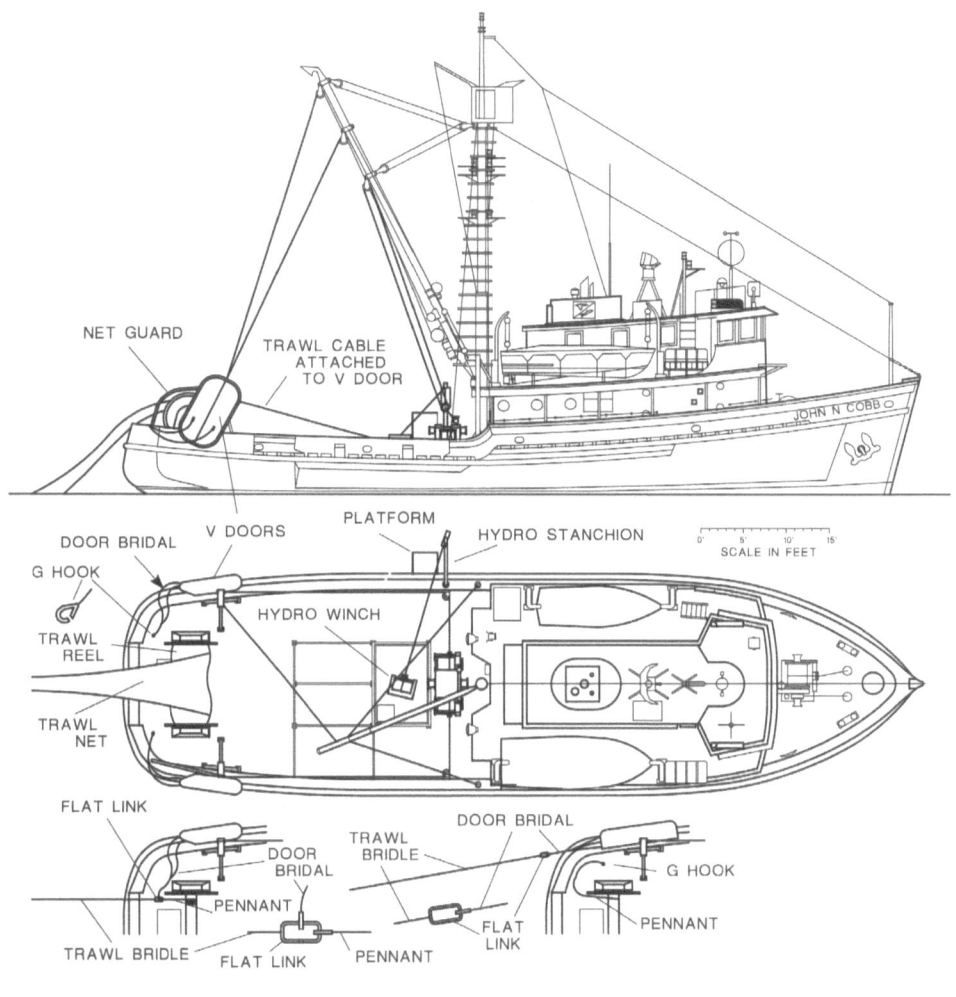

Drawing 14 R/V *John N. Cobb* Modifications (1963).

A serious mishap occurred during the latter part of the first *Commando* cruise (AEC-2) when the stanchion on the port side buckled after hanging up during a tow. This occurred while exploring the northern track line off Destruction Island, the B series north of the Astoria Canyon. As I recall, the damaged stanchion was pulled down to the rail apparently when the net hung up on the bottom, the force involved pulled the stanchion down and bent it in half. It gave me a lot more appreciation for the forces that we were dealing with. I was surprised to be assigned my fourth trip, *Cobb* Cruise 53, especially as the Chief Scientist, which was really a surprise. As my first cruise as CS, I probably was the only one who enthusiastically said yes to the offer of the winter cruise. The AEC work was a great opportunity to observe what species were found at specific depths, especially rockfish, along the continental slope.

Cruise 53 (AEC 5): My Fourth Trip (1962), First Trip as Chief Scientist

We got underway from Pier 92 in Seattle Monday evening, February 26, 1962, after taking on water and fuel oil, and headed for the coast of Oregon, traveling out of the strait at night and down the Washington coast during the day. It was winter, and I was under the eaves on the outside of the pilothouse on the port side during daylight. We were traveling south along the coast of Washington. I hadn't gotten my sea legs. It was a full Bonamine pill day for me. Fresh air and looking for the horizon hidden behind the fast-moving clouds helped control my seasickness and keep me awake. Hopefully, I would get my sea legs soon.

As we proceeded south, I observed a sight I still remember after all these years—the large dorsal fin of a male killer whale that suddenly appeared as it surfed down the face of the swell in front of the *Cobb* and blew out old air and took in fresh air. There were other smaller fins that appeared to be females of a pod with an unknown destination to the east of us. Some of the animals passed forward while others went underneath the *Cobb*. I was impressed by the sight of these large animals at ease in this environment. Once they passed and the routine continued, I found that every day at sea was different. The environment constantly changed from one day to the next, and what life was under the surface was a mystery I was trying to understand. During winter, the weather was always a factor, and the Oregon coast was a place where the only safety you could get was to cross the ocean bar that offered protection in a safe harbor. If you didn't get across before a storm, the Coast Guard would close the bar because of breaking waves, and you would have to either ride it out or go up the coast to the straits and seek refuge in Neah Bay. Unlike the Newport Bar, where we worked during my second trip, the Columbia Bar was much larger and stayed open longer than the others. While we worked the AEC stations, the closest location where we could be protected from the winter storms was to cross over the Columbia Bar and into the river. The Columbia River was a major shipping route, with oceangoing bulk carriers going upriver to take on wheat, which they delivered overseas to India and China. Because many ships have not made it across that bar and have perished in the storms, it has become known as the "Graveyard of the Pacific." Getting into protective waters, such as the town of Astoria, Oregon, just inside the Columbia Bar, and tying up at the city dock was well worth the effort instead of weathering out a winter storm at sea (Fig. 19).

Fig. 19 Crossing the Columbia Bar Before a Winter Storm. Image #4461.

If we got caught in a Pacific coast winter storm, which I hoped I would never experience, the *Cobb* was designed to survive such storms. Pete, the skipper, said, "Once you get a break in the weather, you can get a lot done in a short time, so you want to be as close as you can to your working area so you can get back to it when the weather breaks."

AEC Stations, A Series — Oregon Coast (1962)

AEC Stations are based on depth—every 25 fathoms; and are numbered chronologically shallowest to deepest. Example 3 (100 F) and 4 (125 F) F= fathoms.

We started the survey on February 28, 1962, Day 1, by making a haul at the first station 1-A (50 F) south of the Columbia River. The hauls made at the end of Cruise 50 were converted to the station number based on the depth at every 25-fathom interval, starting at the 50-fathom depth (Drawing 15). The wind was still northeasterly with snow when we set the gear. Joe Dunatov, lead fisherman, noted that the trawl was a western, not eastern, trawl, which I believed we were to use. So we retrieved the net after only a half-hour into the tow, changed to the eastern trawl, and then made a successful hour haul on station 2 (75 F), with a catch of 715 lb. dominated by the yellowtail rockfish (*S. flavidus*). The weather remained choppy with a moderate swell. The weather report called for gale-force winds, so we decided to go back, repeat station 1 (50 F), and then head into Astoria. The drag was successful with a catch of 420 lb., again dominated by yellowtail rockfish.

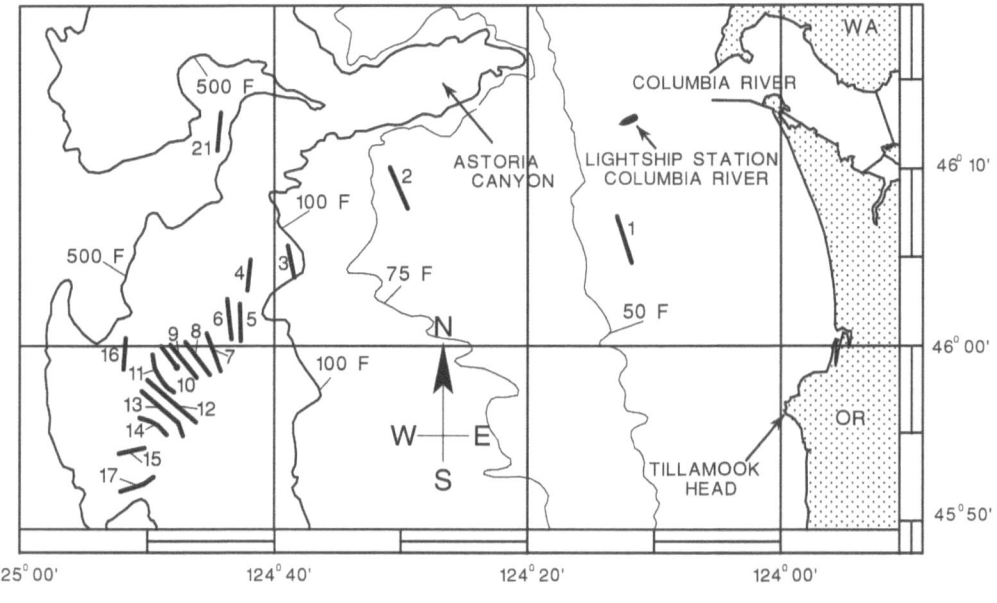

Drawing 15 AEC Stations—Cruises 53 (AEC-5) and 57 (AEC-9).

We pulled into the port dock. While the vessel was being secured, Pete asked me to find the watchman and ask him if it was OK for us to tie up. The *Cobb* had all its deck lights on in the winter's darkness. Looking back at the vessel, I admired her lines and beauty. She was a showpiece, and the white hull all lit up was a sight that sent goosebumps down my spine with pride. I finally found the watchman making his rounds and introduced myself, saying I was off the federal government's research vessel, *John N.*

Cobb. I asked him if it was all right to use his dock off and on while we were in the area. He said, "Yes, you are welcome anytime." Those were the days. I found it an enjoyable task, and it became my job whenever I was on the vessel.

Day 2, we departed Astoria at 3:30 a.m. and got outside to find that the wind was at thirty-five knots, so we turned back in and stayed in Astoria the rest of the day. The weather report in the evening sounded good for the next day, but the lightship reported the wind was blowing forty-five knots at the time.

Day 3, we departed at 5:30 a.m., but when we got out to the lightship, the wind was northwest at twenty-eight knots, and the waves were moderate. We decided it wasn't worth it and turned back to Astoria and the city dock. Day 4, we got underway at 6:00 a.m. and again had a feeling we wouldn't make it, but we found the wind blowing at twenty knots at the lightship and moderate seas. We went on out and made a haul at station 3 (100 F), 4 (125-F), and 5 (150-F). The first one yielded 400 lb. of mainly turbot, dogfish, and skates, but the other two each yielded 2,000 lb. of POP.

It was routine for the CS and the skipper to listen to the Coast Guard radio broadcast in the morning and evening when they gave the weather notice to mariners. We would discuss what the plan for the following twenty-four hours would be. The *Cobb* was drifting near AEC station 6 (175 F) at the end of Day 4. It was still fine weather where we were, but gale-force winds with gusts of forty knots were predicted for the next day. Pete suggested we run into the *Columbia River* lightship station, a two-hour run. This would give us a good chance to cross the bar on a fair tide if it blew up, or to return to the grounds in a short time if it didn't.

On the early morning of Day 5, we were off the lightship. It was blowing at twenty-four knots at 5:00 a.m., and by 5:30 a.m., the wind was picking up to thirty knots, so we decided to go into Astoria. By the time we reached the port docks, it was blowing forty knots. We spent the rest of that day and the following alongside the dock. A wooden vessel is cozy. I could understand why fishermen weren't keen on changing to steel as a construction material. It was warm inside the vessel, with the wind of a winter storm howling outside. The galley was the heart of the vessel. The galley stove was going all the time, fueled by the diesel oil that ran the engines in the engine room. The stove added wonderful smells as the cook prepared the meals and coffee percolated on the stove day or night, enticing the crew. The bunk was the only place one could get any privacy. By pulling the curtains closed, you could cut off most of the world. I often retired to mine while in port waiting out the weather and about to fall asleep, I could hear the wind whistling in the rigging, even over the auxiliary engine that was running constantly, supplying the ship with power. Thank goodness we were in port, tied to the dock. When a violent gust would move the vessel alongside the dock, my thoughts would turn to what it was like on the other side of the bar. What was it like out on the ocean?

One of the two auxiliary diesel engines was always running when the vessel could not hook up to shore power and the main engine was off. When the ship was underway, the main engine, a slow-speed diesel exhaust, produced a pleasing, relaxing tone from the stack on top of the house. During warm weather, I would sit on the starboard side of the bridge with the starboard door open when we were running between stations and listen to the purr of the main engine. When I went down into the engine room, the noise was deafening.[3]

The morning of March 6, Day 7, we departed at 6:00 a.m. with a fair-weather report from the night before. We crossed the bar and headed out to the grounds. Once there, we made a haul at station 15 (400 F) for a total of 710 lb. of sablefish, idiot (*Sebastolobus*), and a trace of Dover sole. Then, at station 17 (450 F), we made another haul for a total of 570 lb. of the same species. They were the deepest hauls we were required to do during this cruise. Afterward, we shut down for the night and drifted. The weather report called for decent weather for the next day.

On the morning of Day 8, we were close to station 16 (425 F), so we made a haul on it after sounding it out. The catch was mainly sablefish, 1,036 lb. Then we moved into the shallower waters, where we had left off before the storm, and made a haul at station 6 (175 F), which produced 1,004 lb. of POP. The third haul for that day was at station 7 (200 F), which yielded small catches of POP and *S. diploproa*. The fourth and last haul for the day was made at station 8 (225 F); the dominant species were sablefish, turbot, and *S. aleutianus*. We shut down for the night and drifted. The weather report was good for the next day. At this point, we had completed the hauls from stations 1 (50 F) to 8 (225 F) on the shallower side and completed stations 15 (400 F) to 17 (450 F) on the deeper side, which left six stations in the middle of the pattern to sample, stations 9 (250 F) to 14 (375 F).

Day 9, we were underway early in the morning. The wind was light, but there were still whitecaps. Once we arrived at station 9 (250 F), we set the gear. The wind was a little more than twenty knots at the time, but an hour later, when we started hauling, it was up to forty knots. Swells were moderate before, but they got large fast, and we had trouble getting the catch in. We had to sort the catch fairly quickly but got a rough estimate. The total catch was estimated at 1,170 lb. of primarily sablefish. Once everything was secured, we were on our way to Astoria. We got across the bar just in time. The weather report for Day 10 was not good, but we took a run out to the bar. There were large swells breaking all the way across the bar, so we returned to the port docks for the rest of the day.

We were underway again the next morning, Day 11, at 2:30 a.m. with a good weather report. Crossing the bar on a good tide, we arrived on the grounds at 7:00 a.m., where we found a large swell and a chop that made it too rough to set the gear. So, we sounded the bottom for the 600-fathom station. The grounds looked very poor and too steep for a tow, but there were possible tows at 500 and 550 fathoms. The morning's weather report was still good, but the swells remained large. We headed for Astoria, arriving there in the evening.

On Day 12, we were on the grounds by 10:30 a.m. after leaving the port docks early. We set the gear on station 10 (275 F) and hung up. It was close to or the same as the one the *Cobb* hung up on when she was establishing the stations on the first AEC cruise. Lead fisherman Joe Dunatov said it would take considerable time to repair the eastern trawl that had been damaged in the tow. There was a western trawl ready to go while the other eastern trawls were in the hold. This was the same one I had understood we were not to use when we made the first set of this trip. I wanted to get in as many hauls that day while the weather held, so we put on the western trawl after I reread the Project Instructions, which stated that a standard 400-mesh otter trawl, not a western or an eastern trawl, so I assumed either would be fine. Pete and I decided to make a haul at station 11 (300 F), which was successful with 996 lb. of black cod. Then we made a haul at station 12 (325 F), which was almost a water haul with a catch of 15 lb., possibly because of the strong tide. We shut down for the night and drifted.

The next two hauls, made the following day, Day 13, were successful. The first one made at station 13 (350 F) had a catch of 860 lb. of sablefish, while the other one, made at station 14 (375 F), had a similar catch—658 lb. of sablefish. The vessel sounded the rest of the day and that night, looking for a possible drag deeper than successful station 17 (450 F), which the *Commando* had established. The grounds deeper than 475 fathoms didn't look good in the area. The skipper finally found a suitable bottom at 550 fathoms in the Astoria Canyon.

The next morning, Day 14, the *Cobb* set the trawl on the possible 550-fathom haul. After towing for an hour, it was retrieved. When the doors reached the surface, they were crossed. The gear was cleared, a dangerous process, and once the doors were separated and placed on their proper stanchions, we could resume fishing. The trawl was being set a second time at the 550-fathom depth interval. The vessel had quite a bit of the trawl cable out when the sounder showed a cliff in front of us that came up 150 fathoms to 400 fathoms. The skipper slowed the vessel, started the winches, and retrieved the gear. Then he reset the gear at the same depth, missed the cliff, and made a successful hour tow, which became station 21 (550 F). The catch was 710 lb., made up of starfish and sea urchins. Out of twenty-three deep-sea metal floats along the head rope, only five survived the 550-fathom tow. The pressure was so great they collapsed as flat as pancakes. They were the same floats Lee had used during Cruise 9, the first deepwater survey made in 1951, and were good to 530 fathoms but not 550. The skipper and I decided to move the operation north to Destruction Island off Washington State and work on the second line of stations for the AEC study, the B series. We had completed all the stations we were assigned, 1 (50 F) to 17 (450 F), and could not find suitable trawling grounds in the deep water for establishing stations 18 (475 F) to 20 (525 F). We located station 21 (550 F), but at that depth, the head rope floats could not hold up to the pressure. We put into Astoria, Oregon, and dropped off the Oregon biologist, who was aboard to tag Dover sole, ending Day 14 and the A series of the AEC track line.

AEC Stations, B Series — Washington Coast (1962)

The *Cobb* departed the next morning, Day 15, and moved up the coast of Washington, arriving at the start of the track line southwest of Destruction Island in the late evening. The skipper and I went over the hauls we hoped we could make the next day. Day 16, we made the first two hauls at stations 1-B (50 F) and 2-B (75 F), which were successful. However, on the third haul, made at station 3-B (100 F), the net hung with no damage to the trawl. We tried another tow in 100 fathoms, but it hung up, damaging the trawl and parting the breast line, so we then rigged an eastern trawl from the hold. We sounded the rest of the day and evening and drifted for the night.

The next morning, Day 17, the wind was up to twenty-two knots. The seas had a good chop, making one of those days you wondered if you should set or not. The vessel continued sounding out another 100-fathom drag, which looked promising, so we set the chain without the net and within ten minutes hung up. As the weather moderated, we moved out to sound the 200-fathom contour, which didn't look good, so we sounded 250, which looked promising with only the chain, but after an hour of towing, the gear came up with crossed doors. The crew and I thought it was too deep for the chain only, so after untangling the doors, we replaced the chain with a trawl and a snag cable stretched in front of the net and set the gear on the preliminary station again. It was successful and became station 9-B (250 F).

The gear was retrieved with a huge catch, an estimated 12,000 lb. of sablefish. They are not buoyed like rockfish, with an expanded air bladder that floats, but are more like lead balloons hanging in the net, waiting to go to the bottom with the net straining to hold the weight. I made a quick decision to let the entire catch go because of the weight. I remember saying to the deck crew, "Let it go." They got the cod end up with somewhat of a struggle because of its weight, untied the puckering string, and let the entire catch out of the cod end without taking a single fish aboard. You could see the catch was entirely small sablefish, and I assumed the crew would be happy with my decision.

Just after the catch was let go, I contacted the office on the base's radio station and told the operator that I had just let 12,000 lb. of sablefish go without taking any aboard. Lee Alverson, who apparently was in the radio room listening, grabbed the mike and chewed me out. My ears burned for days afterward. He told me that was a rare catch and that I needed to take a sample aboard to get a good idea of the consistency of the catch and the size of the fish. He told me I should have taken the first split, the middle split, and the last split aboard, lifting the other splits out of the water alongside the ship and checking the catch as it dropped into the sea to get an idea of what was in it. He further stated loudly that the government was spending a considerable amount of money on the vessel to gain knowledge of the resource, so "do not throw it away." He apparently felt that I was trying to be nice to the crew because then he said, "The crew is being paid to bring the catch aboard for the scientist to analyze." I'm sure word got around to the rest of the crew because big Arnie, the quiet fisherman, came up with a jingle, which he would sing quietly when I was near. "If you want it to go on the fritz, just call Hitz." I learned my lesson. It wouldn't happen again.

The morning of Day 18, we set on a possible station, 10-B (275 F), and retrieved it. The net body was torn up badly, with about 2,100 lb. of fish left in it. So we set out the net again on another possible station, 11-B (300 F). It was also badly torn up, so we decided to run into Neah Bay and mend the nets in order to have a couple to use. We were in Neah Bay all day Sunday, Day 19, and had two trawl nets ready to go on Monday morning.

We departed Neah Bay on Monday, Day 20. The weather had deteriorated. We dragged for butter sole, a special request from the University of British Columbia to collect 150 butter sole. We spent the day trawling along the bread line at various depths to obtain the sample, which was placed in the freezer. We returned to Neah Bay for the night with plans to head back south to the 200-fathom contour and continue looking for stations for the second AEC track line. Day 21, we went out, but the weather was worse than the day before, so we returned to Neah Bay. We tried it again the next day and got near the grounds about 9:30 a.m. with wind increased to thirty-five knots and a large northwesterly swell. The weather report predicted gale-force winds. I consulted with the office via the radio about whether to wait out in Neah Bay for another day or come home a day early. The office personnel gave us permission to come in, ending the first winter AEC cruise on the first day of spring, the day after the equinox, which fishermen say is one of the roughest times of the year.

Cruise 57 (AEC 9): My Sixth Trip (1963), Third Trip as Chief Scientist

The *Commando* was crewed by the same two people that ran the vessel when I conducted my graduate work in 1959 and 1960, the skipper Tom Oswald Jr. and the engineer Olaf Rockness. They had located seven stations in the deep water below 450 fathoms before I went out on my second winter trip on the *Cobb*. The first station was 18 (475 F), which could be sampled with a commercial 400-mesh otter trawl. Deeper stations required a different gear solution to sample them. Both the *Commando* and *Cobb* had duplicate trawl winches and each had two drums, which held 1,000 fathoms (6,000 feet) of half-inch cable each. The otter trawl needs two cables to fish properly as there is not enough cable on the winches to fish much over 550 fathoms. The aluminum floats implode at 550 fathoms.

A decision was made to modify a gulf shrimp trawl that was designed to be fished by a single cable for the double rigged Gulf of Mexico trawler. Each was fished from one of the booms that extended out from each side of the vessel. On the *Commando,* they developed a system using the cable from the starboard winch. Once the 1,000 fathoms of cable was let out, it was attached to the second cable stored on the port winch. This enabled the shrimp trawl to reach the deeper stations from 500 to 1,000 fathoms. Like the garden house, when pulled out from storage, it would kink up after being unwound. The single trawl cable did the same thing after being wound on the winch. The crew of the *Commando* experimented with steel swivels. By fabricating a steel plate to attach the swivels, they solved the problem. Another problem that was solved was establishing a scope ratio to depth of the different types of gear through experimention with the *Commando*.[4]

The eleven aluminum floats used on the otter trawl during my first winter cruise on the *Cobb* had collapsed because of the pressure at 550 fathoms. When the head rope floats collapsed, the head rope also sagged, closing the net. The shrimp trawl could use twenty-two four-inch glass floats along the head rope, which all held up in water depths of over 1,000 fathoms. There was a lot of experimentation done on the *Commando* in developing a sampling device in the deeper waters where the otter trawl could not be used.[5] Drawing 16 is a sketch of the two different trawlers along with the scope ratios. The information on the 400-mesh otter trawl is from Greenwood's 1958 publication and the Gulf shrimp trawl from his 1959 publication.[6,7]

Sometime between August 23 and September 6, 1962, I went out on the *Commando* summer cruise (AEC-7) to observe the technique they had developed to sample the deep waters between 600 and 1,000 fathoms. The shrimp trawl they used had two V-doors, each attached to a wing of the net. Both doors were attached to the end of the starboard trawl cable through a steel plate with steel swivels. The steel plate was pulled up to the starboard stanchion with the net hanging directly behind the starboard side of the vessel once it was put over the side. The starboard drum brake was released until the doors were underwater, then the break was used to stop the cable from playing out. If the doors were observed to spread out evenly to opposite sides, the amount of cable was let out that was calculated from the scope ratio table for the depth of the station. Once the depth required

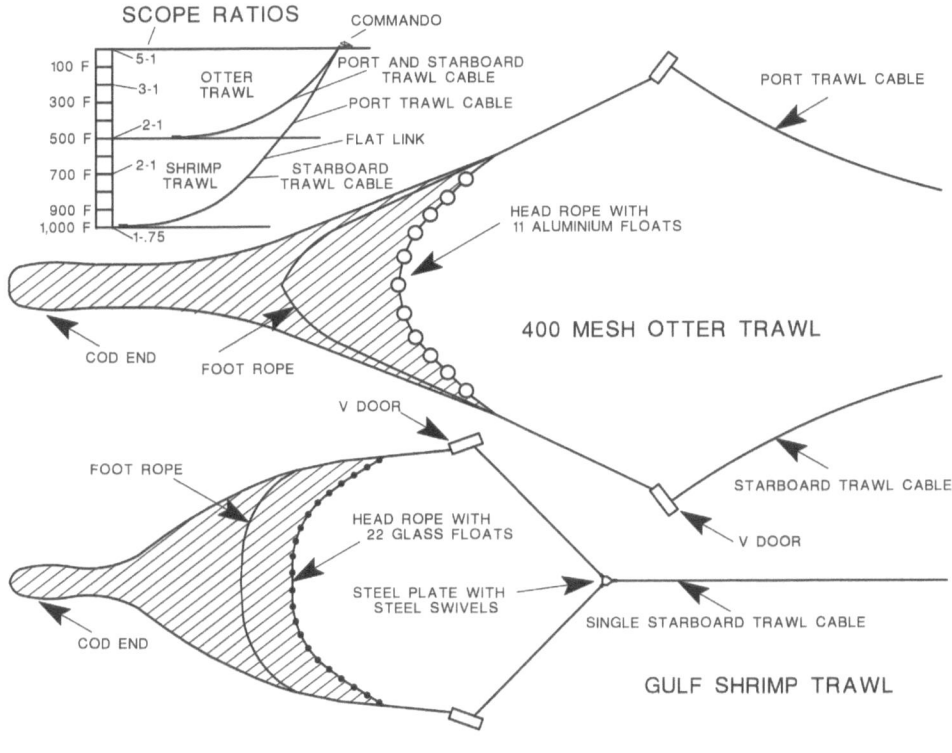

Drawling 16 400-Mesh Otter Trawl, Gulf Shrimp Trawl.

was more than 100 fathoms of trawl cable stored on the starboard drum, it was stopped when the flat link reached the starboard stanchion and went through the block. The end of the port side wire, lying slack on the deck, was taken over and hooked into the link, which was under considerable tension between the deployed net and the entire trawl cable in the water from the starboard winch. By slacking off the starboard winch and hauling in on the port one, the tension was transferred to the port side. Once the entire strain was taken up by the port cable, the slackened starboard cable was released from the flat link that was attached to the port cable. The correct amount of cable, determined by the scope ratio was let out. To see the net returned with a catch in the cod end was rewarding.

The scope ratio (length of cable to depth) was used by trawl fishermen to determine the amount of cable to use for a trawl haul. A ratio of three to one was used for hauls made at 50 fathoms. From 100 fathoms to 500 fathoms, they used a ratio of two to one. Deeper than 500 fathoms, the crew of the *Commando* had to develop their own ratio, which came out to be close to two to one in water depths out to 800 fathoms and a ratio of one to one in the deepest station of 900 to 1,000 fathoms.

On January 21, 1963, I was on the second winter *Cobb* AEC cruise as the CS. We had a streak of good weather at the beginning of the trip up to when we finished occupying the stations from the deepest station, 17 (450 F), working into the shallowest station, 1 (50 F). On Sunday, January 27, 1963, we completed all of them in five continuous days. The catches made during the two winter cruises were all made with the otter trawl, not the shrimp trawl, and were on the light side. In 1963, *Cobb* Cruise 57 only had one haul over 1,000 lb., which was 1,426 lb. of POP at station 5 (150 F), whereas in 1962, *Cobb* Cruise 53 had four stations with catches over 1,000 lb. Three of them had catches of POP that ranged from 1,004 to 1,600 lb. at stations 4 (125 F) to 6 (175 F). The other large catch was 1,036 lb. of sablefish taken at station 16 (425 F). The year prior had only one station with a catch over 1,000 lb.—station 5 (150 F) with 1,426 lb. of POP.

These catches reinforced my understanding of how different species occupied certain depths along the bottom of the continental shelf and slope. There were basically three species found on the shelf that extended to the slope in almost all the stations out to station 21 (550 F)—Dover sole, sablefish, and idiots (*Sebastolobus alascanus*). Three species of black rockfish were found near the continental break—*S. brevisipinis, S. flavidus,* and *S. entomelas.* There were twelve species of red rockfish taken. *S. aleutianus* and *S. aurora* were found in the deepest haul, and the latter was a new species I had not seen before. Three species—POP (*S. alutus*), *S. crameri,* and *S. diploproa*—were found in all four stations from 4 (125 F) to 8 (225 F) during both years. The rest of the rockfish—*S. elongatus, S. helvomaculatus, S. paucispinis, S. pinniger, S. proriger, S. rubrivinctus, S. saxicola,* and *S. zacentrus*—remained in the shallower side of station 6 (175 F). No rockfish were found deeper than station 9 (250 F).

The *Commando* established station 18 (475 F) with the otter trawl and six deeper stations 500 to 1,000 fathoms with the shrimp trawl stations 19-S to 24-S (S=shrimp trawl) (Drawing 17). A black circle indicates the location of the shrimp hauls, and a dark line, the otter-trawl hauls. We were to duplicate each of the shrimp hauls along with adding a new station to *Cobb,* haul 18, which the *Commando* had established. The new one was with the otter trawl and the rest with the shrimp trawl. The latter were found in areas where the shrimp trawl could be towed, which were

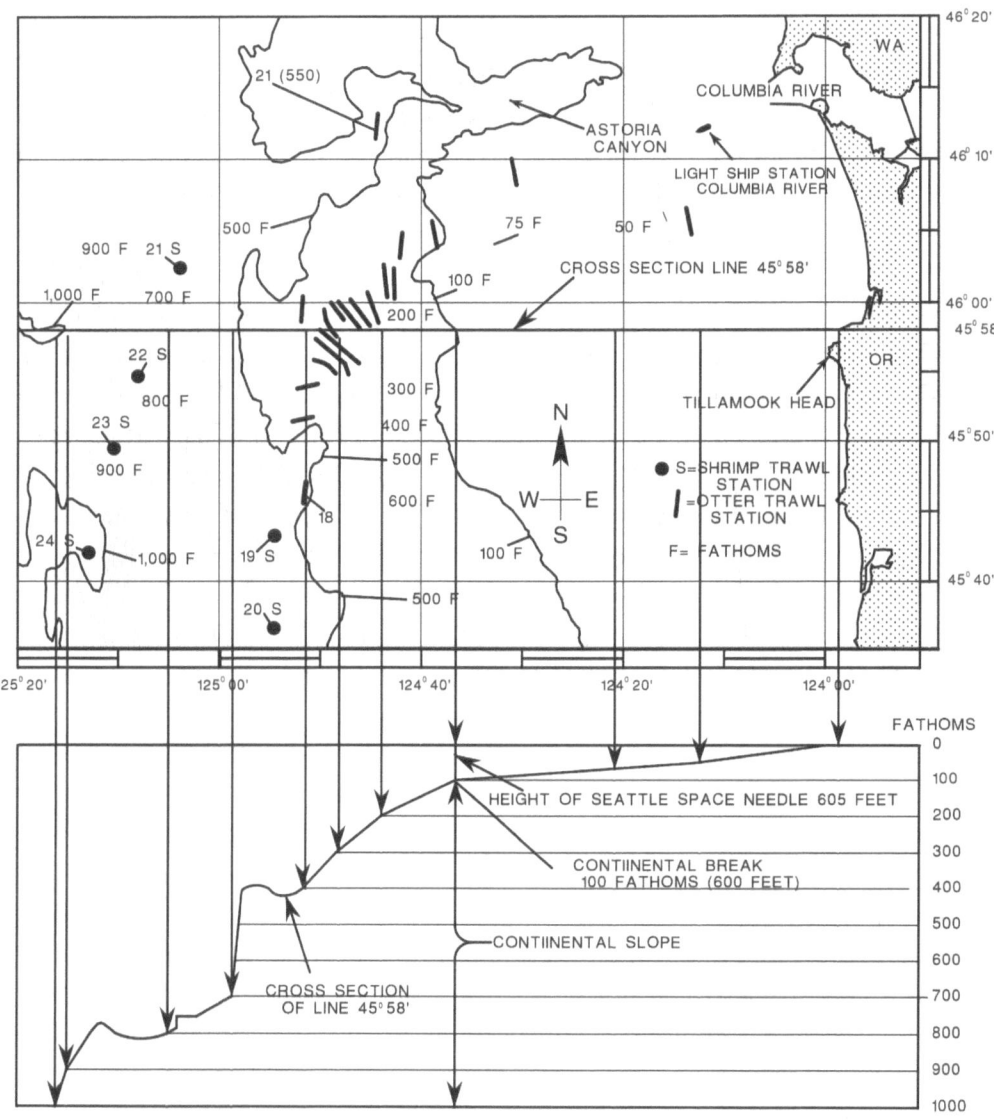

Drawing 17 Deep Water AEC Stations—Cruise 57 (AEC-9).

difficult to find. The bottom of the ocean in the depths below 500 fathoms became steep with cliffs and small peaks. There was a major change in the numbering of the stations. A decision was made in the remaining six stations that were sampled with a shrimp trawl to space them 100 fathoms apart instead of 25, starting at 500 fathoms.

The cross section of line forty-five degrees fifty-eight minutes is projected vertically by depth, giving the reader a feel for how fast the slope drops off to 1,000 fathoms. The shelf is relatively flat, whereas the continental shelf becomes quite steep. There is enough flat space in the upper part of the slope to establish the AEC stations out to 475 fathoms. Then the cross section shows that there is a cliff from 400 to 700 fathoms. The *Cobb* made a haul in the Astoria Canyon the winter before when they were still trying to establish AEC station 21. During the first try, they were heading for a cliff. Once the vessel retrieved the trawl and changed course, the next haul was successful, but the head rope floats could not stand the pressure and imploded. The *Commando* explored for the deepwater tows with the shrimp trawl and found six of them. The deepest one was the AEC station at 1,000 fathoms, referred to as number 24-S. It wasn't easy keeping the trawl steady so it would not spin and close the net or mud up. The technique that the *Commando* developed in sampling the deep stations with a shrimp trawl took a lot of time setting the gear and retrieving it.

I reviewed the *Cobb's* deck logs for the sixteen days after we completed the assigned stations using the otter trawl on Sunday, January 27, 1963.[8] The logs helped me recall what happened during that time. The next morning, Monday, we set out for station 18 (475 F), which the *Commando* had located. We were to use the commercial trawl on it. When we reached the lightship station *Columbia River*, the wind was northeast at Force 6 or 7, with rain and snow. The *Cobb's* log had a record of the weather conditions every two hours, but they were expressed using the Beaufort Scale.[9] Force 7 lists the wind speed at twenty-eight to thirty-three knots, strong wind, seas 13.5 feet high, and large waves, streaky foam.

We finally turned back to Astoria and tied up at the boat basin, using the time to set up for sampling the deeper stations using the shrimp trawl. Since it was the first time the *Cobb* used the shrimp trawl, a pendant had to be added to the starboard winch. All the cable was pulled off the starboard winch, then a 20-fathom pennant was wound onto the winch. The end of the main trawl cable was attached to the pennant through a detectable link, or a flat link. All of it was wound back onto the starboard drum. The shrimp trawl and doors were attached to the starboard trawl cable just aft of the starboard stanchion. We were ready to sample the deepwater stations.

We departed the upper boat basin the next morning. The weather was clear, winds east at Force 2, but then increased east-northeast at Force 6, so we turned around and headed back to Astoria. By the time we arrived, the wind had increased to Force 7. We tied up at the port docks and remained there for the next two days, while the winter storm winds caused the mast rigging to whistle. I remember an incident that occurred when we were tied up at the city port docks during this time. Snow was blowing across the dock and the *Cobb*. I was warm and cozy in the pilothouse with Pete, the skipper, waiting for the morning weather report from the Coast Guard. We watched a man come down the lee side of the warehouse and walk around the corner. The dock was slippery with snow and ice. He was about an eighth of the way to the door he was headed for when a winter gust hit him and knocked him down. He crawled back around to the lee of the building where he came from. I was just thankful we were inside and not outside of the bar in the winter storm.

On Thursday night, the weather report predicted gale warnings along the Oregon coast for south to southwest, with twenty-five to forty-five knot winds during the night, becoming fifteen to thirty-five knots Friday and Saturday. So, Friday, February 1, we departed Astoria at six in the morning. We reached the site of the deepest station, 24-S (1000 F), and decided to set the haul. This was the first set of the shrimp trawl that was towed by a single cable that the *Commando* had developed. All the cable from the starboard winch was let out and then transferred to the cable from the port winch. Once the port cable took the strain, it was disconnected from the starboard cable. The port side cable was let out to the desired amount before the port winch was stopped. After the shrimp trawl was down on the bottom for an hour, it was hauled back. Once the cable was wound on the port winch drum and the connective link between the port and starboard cable arrived, it was stopped just aft of the port stanchion. The port trawl cable was under tremendous tension and was transferred from the port to starboard stanchion by connecting the slack starboard pennant cable into the connective link and then engaging the starboard winch and releasing the port one. Once the tension was taken up by the starboard stanchion, the port cable became slack and was disconnected and the remaining 1,000 fathoms of cable were retrieved. I was really excited to see what was in the catch from such a deep haul but was depressed to see the bridle and doors all wrapped up with crossed doors closing off the net, so there was nothing in the cod end. By the time the net and doors were untangled, the vessel was shut down and drifted for the night.

The next morning, the gale warnings were posted, so we headed for Astoria. The wind came up to Force 7, and we got across the bar into Astoria just in time. The next day, gale warnings were still posted, so we moved the vessel upriver to the Union Oil dock, which was inside a breakwater. That night, because of the strong tides, a deck watch was set up to manage the lines that held us to the float. I had the 8:00 pm. to 10:00 p.m. watch; Pete Larsen had the next one, 10:00 p.m. to midnight. The lightship was blown off station that night. Just before midnight, there was a shower of sparks and then another. A huge barge had broken away upriver and came drifting down, crashing into the breakwater several times. As it bounced off, it caused showers of sparks. It was huge. The bow hung over the breakwater, but finally grounded on a sandbank downriver.

The gale warning remained up, the lightship remained off station, and the Columbia River pilot boat came in. Two large ships anchored upriver from us off Tongue Point were waiting to get out. The Astoria airport had winds of sixty-five knots, and the water on the river was smoking—that is, the wind gusts would pick up the water off the surface, which appeared to be smoke.

Monday, things changed. The four large ships anchored upriver started downriver, and another three followed. When the last one passed us, we set out and followed them out across the bar. We passed the huge barge that broke away and rammed the breakwater. It was stuck on the sandbar. We passed four incoming ships and a large tug and barge heading upriver, crossed the bar, and headed north. It was an interesting trip up the coast, with the following seas pushing us along until we reached the straits, where we turned west and headed for home. I was disappointed at not sampling the hauls the *Commando* had located with the modified shrimp trawl in the deep waters of the AEC track line. I swear I could hear Arnie singing his little jingle as we headed home. The closure of this trip was my final trip with the AEC project, but the catches clearly demonstrated that species live at certain depths below the ocean surface.

Al Pruter and Lee Alverson edited a book on the AEC project, published in 1972, *The Columbia River Estuary and Adjacent Ocean Waters: Bioenvironmental Studies*. Miles Alton wrote chapter 24, "Characteristics of the Demersal Fish Fauna Inhabiting the Outer Continental Shelf and Slope off the Northern Oregon Coast."[10] He had all the catch information taken during the AEC project and went into detail about what species were taken and at what depth. He concluded that the largest accumulations of fish were obtained from the 50 to 225 fathom intervals. Important components of the catches were Pacific hake (*Merluccius productus*) at 50 to 175 fathoms, flounders at 50 to 225 fathoms, and rockfish, mainly POP (*S. alutus*), at 100 to 225 fathoms. In the latter group, there were approximately sixteen other species of *Sebastes* taken in the hauls. A secondary zone of abundance was 250 to 425 fathoms, where sablefish (*Anoplopoma fimbria*) comprised most of the fish caught there. In this same zone, the idiot, shortspine thornyhead (*Sebastolobus alascanus)*, was found out to 650 fathoms and was replaced by the longspine thornyhead (*Sebastolobus altivelis)* and extended to a depth of 850 fathoms almost to the abyssal plain. He presented the information on the species that occurred on the shelf and the slope and the deepest depths that I observed in my book was 425 fathoms.

There were seventeen species of *Sebastes* taken during the three year AEC surveys. They were *S. aleutianus, S. alutus, S. aurora, S. brevispinis, S. crameri, S. diploproa, S. elongates, S. entomelas, S. flavidus, S. helvomaculatus, S. paucispinis, S. pinniger, S. proriger, S. ruberrrimus, S. rubrivinctus, S. saxicola,* and *S. zacentrus*. POP was the most important rockfish species taken during the AEC survey, with catches taken between 125 and 175 fathoms frequently greater than 1,000 lb. per hour. The greatest catch of *S. alutus* was 7,000 lb., made at the 125-fathom station. *S. crameri* ranked second to *S. alutus* in catches taken on the slope at stations from 125 to 175 fathoms. With the closure of this trip, I ended my participation with the AEC work and was assigned a new task.

Chapter 9: Endnotes

1. A. T. Pruter and D. L. Alverson, eds., *The Columbia River Estuary and Adjacent Ocean Waters: Bioenvironmental Studies* (Seattle: University of Washington Press, 1972).

2. H. Heyamoto, "Techniques and Equipment Used by the Bureau of Commercial Fisheries," in *The Columbia River Estuary*, ed. A. T. Pruter and D. L. Alverson, pp. 395–407.

3. Enter "*John N. Cobb*—1950 Fairbanks-Morse Diesel Engine" into search browser for a video on the sound of the exhaust and see the engine room on YouTube. https://www.youtube.com/watch?v=-2ksTQGTlLA.

4. Walter T. Pereyra, "Scope Ratio-Depth Relationships for Beam Trawl, Shrimp Trawl, and Otter Trawl," Commercial Fisheries Review, vol. 25, no. 12, December 1963. MS # 98.

5. W. T. Pereyra, H. Heyamoto, and R. R. Simpson, "Relative Catching Efficiency of a 70-Foot Semiballoon Shrimp Trawl and a 94-Foot Eastern Fish Trawl," *Fishery Industrial Research* 4, no. 1 (November 1967). MS #142.

6. Melvin R. Greenwood, "Bottom Trawling Explorations off Southeastern Alaska, 1956–1957," *Commercial Fisheries Review* 20, no. 12 (December 1958). MS #39.

7. Melvin R. Greenwood, "Shrimp Exploration in Central Alaskan Waters by John N. Cobb, July–August 1958," *Commercial Fisheries Review* 21, no. 7 (1959). MS #46.

8. "*John N. Cobb* Deck Log Year 1963, Month January," Records of the *John N. Cobb*, Historical Ship Files, ID 119654305, US National Archives Seattle.

9. Peter Kemp, ed., *The Oxford Companion to Ships and the Sea* (Oxford: Oxford University Press, 1988), pp. 7. Beaufort scale used to indicate the force of the wind.

10. Miles S. Alton, "Characteristics of the Demersal Fish Fauna Inhabiting the Outer Continental Shelf and Slope off the Northern Oregon Coast," in *The Columbia River Estuary*, ed. A. T. Pruter and D. L. Alverson, pp. 583–634.

CHAPTER TEN

Pacific Hake
1964–1966

During my career at the exploratory base in Seattle, I was assigned a number of different cruises, the last one occurring in 1971. I conducted a number of different projects during these cruises, such as sablefish trap experiments, exploring for scallops, using a hydraulic dredge to locate clams, and hake research. From this point in my book, I will refer to a cruise that relates to my story of POP.

Dick McNeely was transferred to the exploratory group in the 1950s to develop a large slow-speed trawl and a system to control the depth where it fished.[1] He fabricated new hydrofoil trawl doors, which looked like an airplane wing. They were held up in a horizontal position when towed through the water. Water flowing over the airfoils helped pull or shear a door to the side of the vessel, and two doors opposite one another would spread or open the net. He purchased cables with electronic wires running through the center, replacing the conventional trawl cable. He developed a method of diverting the electric current through the trawl winch so that the electrical current from the trawl cable could get to the pilothouse. He then set up a way to pass the current from the trawl cable to a depth-sensing instrument. Because of the stress of the trawl doors pulling on the trawl cable, this wasn't a simple matter. If any saltwater got into the electrical system, it would short out. I remember the O-rings and the epoxy he used to make the connection. The smaller trawl blocks the cable passed over were changed to larger ones so the electrical wires in the cable wouldn't be crushed.[2] It worked, and the Exploratory Fishing and Gear Research Base finally had a workable midwater trawl.

So, in early 1960, the net became a sampling tool, and by the mid-60s, it was used to find and map out North Pacific hake (*Merluccius productus*) (Fig. 20). If a market could be found or developed for this resource, such as fish flour, it would put the herring and sardine seiners back to work, which had been idle since the collapse of the sardine fishery.

Fig. 20 Pacific Hake—*Merluccius productus*. Image #20688.

Cruise 64—My Ninth Cruise (1964)

I got involved with hake when the office assigned me as a CS on Cruise 64, conducted between February 5 and March 20, 1964. This assignment was a trip south to explore the midwater off Southern California and Mexico for Pacific hake. I had a choice between the first or second half of the cruise and decided to take the second half because I thought the prevailing wind off Washington and Oregon was from the south and could sail back in comfort. I flew from Seattle to San Diego, California, arriving on February 28, 1964, and took over as CS.

San Diego was a navy town with palm trees and warm sunshine. I could walk uptown and have a glass of fresh squeezed orange juice. The vessel was moored in front of two large navy ships that dwarfed the *Cobb*. The only way out was to back the *Cobb* out between them. Pete blew the whistle to let any passing vessel know he was backing out, but when he released the whistle's handle, it just kept blowing. The engineer came running up from below to shut it off. Personnel from the two navy ships came running out to see what was going on since the whistle sounded like a big ship, and it kept blowing. They looked out and didn't see a huge ship, so they looked down.

What they saw was the little *Cobb* backing out, I swear I could hear Arnie singing his little jingle, indicating the return of Hitz.

The ocean was so much different from the trips I had made to Alaska, Canada, Washington, and Oregon. Flying fish would break the surface and sail off as the *Cobb* moved through the water. Pelicans with their pouch beneath their bills flew about as numerous as seagulls. At night, you could see porpoises tearing through the water like torpedoes heading for the bow, all lit up in phosphorescence. During the day, they were close, playing in our bow wake. I would hang an orange on a string over the bow and swing it side to side just above the water as we moved forward, and they would roll onto their sides to see what it was.

The objective of the cruise was to determine the distribution and abundance of adult hake during a predicted period of peak spawning because of the work done by the US Fish and Wildlife Service's La Jolla Biological Laboratory. Their R/V *Black Douglas,* working off the California coast, found that February to March was the peak season of hake spawning.[3] They used plankton nets to catch the hake eggs and larvae near the ocean surface. Based on the distribution of eggs and larvae over time, they determined the spawning time. We were there to see if we could find the spawning population of the adult fish. By using the sounder, the spawning population could be located in midwater. Severe weather encountered during most of the cruise limited the midwater hauls to thirty-five. Adult hake in amounts up to 350 lb. per hour tow were taken in five of the drags. Two of the hauls made a fair showing on the sounder during the latter part of the cruise, producing 300 and 150 lb. of hake. Their concentration centered at 250 fathoms below the surface in midwater, and the signs disappeared in the evening hours. Attempts to relocate the shoal on the following day proved fruitless. A correlation was found between the catches made by the *Cobb* and the sampling by the *Black Douglas* with plankton nets that had good catches of hake eggs and larvae above the location where the *Cobb* made the midwater tows.

We got an emergency call from a vessel in distress on the night of March 10, 1964. *White Star*, a purse seiner, was sinking and could not keep up with the incoming seawater. We proceeded to the area where she was anchored close to the northeast corner of Cedros Island, Mexico, close to shore. The image of this large, beautiful vessel is still etched in my mind. The floodlights of the *White Star* lit her up from the mast down in the jet-black night. She was close to the island, and her lights picked up the island's cliff rising from the sea. The vessel was low in the water and would rise and fall with a slight swell. Pete put me on the radio, which was in the after part of the chartroom, with the approaching search and rescue amphibious plane—a US Coast Guard Albatross, a larger version of the Grumman Widgeon—coming from San Diego. I told the pilot which boat we were and the location of the seiner. They flew low over us a couple of times and then dropped a pump that landed next to the seiner, which they retrieved. They got it running and were able to keep up with the incoming inflow of seawater, and we were excused from the sight.

Running back up the coast on the way home, we'd get a drawing from the weather fax that showed the location of the high and low pressure cells as they were moving in the area of the Pacific where we were. There was a huge low, a hurricane off Mexico, moving up the coast behind us. I asked Pete whether we should be concerned with its movement. He said we were off San Diego and far enough north and that they never got this far up. I kept watching it, though, as we moved farther north. We were bucking into the wind and waves. It slowly kept getting worse as we continued north. When we were off San Francisco and could see the lights of the city, Pete asked if we should take a break or keep going. I said keep going; it should get better. About that time, a cruise ship came out under the Golden Gate Bridge and crossed our bow, all lit up. We could hear the music coming from her. They turned south, and we kept going north. How nice it would have been if we were on that ship.

When we got up to the Columbia River, the wind had increased even more. There was a cross pattern of two or three ground swells, plus a large chop stirred up by the north wind. The wind increased again. It was miserable. Where were the prevailing southerly winds I expected? There was no place you could get comfortable on the *Cobb*. If you got into your bunk, you started to get bed sores from the violent movement. If you wedged yourself in a corner, for example, between the closet in the after stateroom and the bulkhead, the movement was so erratic you expected the vessel to move one way, but it would go a completely different way. I accumulated several bruises on that trip.

Finally, we reached the straits and headed in. What a relief! I had survived my greatest fear of becoming violently seasick if we bucked more than a few days. We had bucked for about five days, each day worse than the day before, but I was never seasick. I became less concerned about seasickness. It was just part of the job.

Cruise 67 — My Eleventh Cruise (1964)

Our office was getting more involved in the hake resource and wanted to convert the sampling midwater trawl into a commercial trawl. In November 1963, the gear unit began to charter their own vessels instead of using the *Cobb* for their work. For Gear Cruise 2, which was integrated into the list of exploratory fish cruises as 65A, they chartered the *St. Michael* from May 4 to July 7, 1964, and used the sounder to locate an abundance of Pacific hake that existed off the Washington coast near Destruction Island. Several midwater hauls made on good soundings of fish consistently produced catches of 30,000 lb. of hake per hour haul. These results indicated that the hake resource might be used for the development of fish flour. There was a need to obtain information as to the seasonal availability and abundance of the species.

My next opportunity to be involved with hake was Cruise 67, scheduled for August 10 to October 9, 1964. I joined the vessel during the first half of the cruise. Our mission was to locate hake concentrations and map them out by using an echo sounder, the same one used to determine whether the bottom was hard or soft when we were exploring for new trawl hauls. Now, we were looking for the fish signs, which were separated from the bottom by a white line between them and the bottom. The vessel was run onshore and offshore along transects that bisected the depth contours at oblique angles of 20 to 30 degrees. Transects were drawn on the navigational chart. Once fish signs were seen on the sounder, they were classified into three groups: light, medium, and heavy. The point where the classification of the signs changed was noted on the chart as the vessel proceeded along the track line. Once the signs became light and disappeared, the vessel continued for approximately one mile and then changed course by 150 degrees. If the signs reappeared, they were recorded on the chart as the vessel traveled along the new transect and continued changing course time after time as long as the signs reappeared. Once the signs disappeared, the outline of the school of fish was drawn on the chart. The vessel returned to the area of the school where the signs were heavy.

A midwater trawl was made in the heavy part of the school to determine what species of fish the sounder was showing and in what quantity. The catches were always pure hake with a very small amount of yellowtail rockfish (*S. flavidus*) mixed in. Catches of hake made with the *Cobb* pelagic trawls in heavy or medium traces ranged from 10,000 to 60,000 lb. per hour of towing (Fig. 21).

Fig. 21 Large Catch of Hake. Image #11067.

It was quite a sight to see that cod end appear behind the vessel. With all the air bladders expanded, the bag slowly rose and floated on the surface. When the cod end of the net was alongside the vessel, the first split was brought aboard and placed in the checker. Each split that followed was lifted out of the water alongside the vessel, the weight was estimated, and the catch was released to the sea. As it fell, any fish that was not a hake was noted. When the CS estimated that half of the catch had been released, the next split was taken aboard and placed in the checker. Then the rest of the splits were handled in the same manner as those that were taken before the middle catch was brought aboard. Finally, the cod end was brought aboard with the last of the catch.

On one of our transits during a hake cruise, while looking for traces of hake on the sounder, we saw something that gave me the willies. It was fine weather. You could see for miles. There was a moderate ocean swell and very little wind chop. We were south of the Columbia River, and I was on the bridge watching the sounder. I would often look out of the pilothouse windows, and on the horizon, I saw a black spot that disappeared. Then I saw it again, only for it to disappear again. So I asked Pete if he saw it, and he did. He didn't know what it was either, so we went to investigate.

We broke off the transit and headed for it. As we drew near, we could see that it would disappear completely for over a minute and then reappear. It continued doing the same routine over and over. We could see it slowly go underwater, where it remained a moment before suddenly breaking the surface of the ocean. As we got closer, we saw that it was the stump of a large upside-down fir tree that had become waterlogged. Apparently, the ocean swells caused the tree to rise and sink over and over again. It probably came out of the Columbia River and had taken the upside-down position with time. It gave me the willies.

Cruise 71—My Twelfth Cruise (1965)

I was aboard the R/V *John N. Cobb* in May 1965, and we were mapping out schools of hake, trying to determine when and where they appeared off the coasts of Washington and Oregon. As we made echo-sounding runs across the continental shelf, approaching the 100-fathom contour, we observed a rare sight—a Soviet side trawler. She was slowly working her way south along the continental edge near the area we had explored. It was the first sighting of a Russian fishing vessel in this area that I was aware of.

I knew they were fishing flatfish in the Bering Sea in August 1959 because when the *Cobb* transited through to the Chukchi Sea, it encountered a fleet of Russian vessels. Lee Alverson was aboard and observed the fleet and gave a talk at the University of Washington College of Fisheries when he returned.[4] Lee had a stimulating way of presenting a topic, which was part of the reason I wanted to work for him at the Exploratory Fishing and Gear Research Base. The lecture was published in *Western Fisheries* in April 1960. When the *Cobb* was exploring the Gulf of Alaska in 1962, she again came across two Soviet side trawlers fishing POP along the continental break at the 100-fathom contour.[5]

This was my first encounter with a Soviet vessel off the Washington and Oregon coasts, so the skipper and I decided to observer her a little closer. The *Cobb* headed to her position, just south of Grays Canyon off the Washington coast. As we approached, she turned to starboard (right) and was coming back toward us, making a circle as she started setting her net, a standard procedure for a side trawler setting her gear. As she came out of the circle and was lining up on her drag, she followed the 200-fathom contour, dragging the net along the ocean bottom.

The *Cobb* followed her during the tow, which lasted an hour. At the end, she turned starboard across the mouth of the net. The ship drifted as the winches retrieved the trawl. Once alongside the ship, the cod end with the catch was pulled from the water and dumped into a checker on deck. The catch was small and couldn't be observed, but we believed it was composed of rockfish, sablefish, and Dover sole, which, from past exploratory drags, were the most likely species present along the 200-fathom contour.

Interestingly, we found the drag they made would have snagged our net. They were using bobbins or rollers, which got their net across the bottom without snagging, and had a successful haul. When we got back to Seattle, there was enough interest in the sighting that I published a note in the June 1965 issue of the *Fishermen's News,* "Observations of a Russian Trawler."[6]

Fig. 22 Soviet R/V *Adler,* Author's photo.

The vessel's name was in Russian and translated to *Adler,* and her home port was Vladivostok, Russia (Fig. 22). I became interested in this Soviet trawler. When I got back, I found in the literature that it was built in England as the pioneer class distant water trawler. There were twenty of them built for the Soviets between 1956 and 1958. They each had a total length of 190 feet. The Soviets had apparently decided to harvest the ocean for food for their people. What better way than to purchase the best ships the world produced? The *Adler* was an example.

I assumed that since we were in the Cold War with the Soviets, they were using the *Adler* to check their charts out for their navy while pretending to explore for fish populations. I knew that US Navy planes periodically flew over the vessels working or transiting along the coast since, once in a while, they would buzz us during the day or night. I also believed they were photographing them because, at night, they would fly over us, bathing us in light. Flashbulbs built in the plane lit us up as they took a picture. I assumed they were looking for foreign vessels and must have been monitoring the *Adler*. The Russian crews were friendly when the *Cobb* first sighted them in the Bering Sea and the Gulf of Alaska. As we drifted near the *Adler* on that May day and she began to get underway, one of the winch operators heading into the house raised his arm with his thumb up, a good sign in their culture and in ours. As we left the Russian vessel, she apparently headed south toward Oregon, exploring as she went, probably looking for concentrations of POP, sounding for their navy, or both. I wonder if she found the perch spot off Oregon.

By May 1965, the annual pattern of the hake was understood. Between February 1 and March 12, 1965, the scientist aboard the *Cobb* found schools of spawning hake off Southern California. They were in midwater at 125 fathoms below the ocean's surface, above bottom depths of 500 to 800 fathoms. The schools could be located the next day, moving slowly to the north. On the way south, echo transects were made offshore and inshore along the coasts of Washington and Oregon, but no signs of hake were observed. From March 29 to May 28, 1965, they found no hake off Washington and Oregon during the first part of the cruise but did during the second part. Then, between July and the middle of August 1965, the scientist found hake schools off Oregon and Washington.

By this time, the pattern of the hake was finally understood. Spawning occurred in February and March off California and Mexico, then the adults moved north and arrived off Oregon in the spring, continued moving into Washington during late spring and early summer, and began to disperse in the fall. The schools were large along the continental shelf and could be taken in huge catches with the use of a midwater trawl.

The gear unit chartered a commercial vessel, the *Western Flyer,* in July 1965 to determine whether it was profitable for a fishing vessel to catch and deliver hake commercially to a shore side plant using the midwater trawl. This work was carried out in Exploratory Cruise 72A, as Gear Cruise 6. Based on the results, it was profitable for commercial fishing vessels to do the work. Fish protein concentrate (FPC), developed in 1962, when dissolved in water, was a tasteless high-protein product for human consumption.[7] The federal government was interested in building a plant that could produce FPC near the unused hake resources to help find a solution to world hunger. A new fishmeal plant operated by Pacific Protein Corporation was built in Aberdeen, Washington, in 1965. Five commercial fishing vessels were contracted: *New St. Joseph, San Vito, Baron, Junior,* and *Recruit.* They were supplied with the government-designed *Cobb* midwater trawl as well as the specialized trawl cable that would be used to catch hake in midwater. They would go out of Grays Harbor, find the hake schools on the continental shelf, and catch pure hake. After loading the hake onto their vessel, they would deliver them to the Pacific Protein Plant in Aberdeen. They started catching hake and delivering them to the plant in the spring of 1966.[8]

Chapter 10: Endnotes

1. Richard L. McNeely, "Development of the *John N. Cobb* Pelagic Trawl—A Progres Report," *Commercial Fisheries Review* 25, no. 7 (July 1963). MS #90.

2. Richard L. McNeely, "A practical Depth Telemeter for Midwater Trawls," *Commercial Fisheries Review* 20, no. 9 (1958). MS #36. See also Sep. No. 522.

3. Martin O. Nelson and Herbert A. Larkins, "Distribution and Biology of Pacific Hake: A Synopsis," *Pacific Hake, Fish and Wildlife Circular* 332 (1970), pp. 23–33.

4. Dayton L. Alverson, "The Japanese and Russian Trawl Fishery in the Bering Sea," *Western Fisheries* (April 1960), pp.12–14, 30–31. MS #53.

5. A. T. Pruter, "Soviet Trawlers Observed in the Gulf of Alaska," *Commercial Fisheries Review* 24, no. 9 (1962).

6. Charles R. Hitz, "Observations of a Russian Trawler," *Fisherman's News* 21, no. 11 (1965), pp. 9. MS. #136.

7. Keven M. Bailey, *Billion-Dollar Fish: The Untold Story of Alaska Pollock* (Chicago: University of Chicago Press, 2013), pp. 40. Fish protein concentrate (FPC).

8. Martin O. Nelson, "Pacific Hake Fishery in Washington and Oregon Coastal Waters," *Pacific Hake* (March 1970), pp. 43–52.

11
CHAPTER ELEVEN

Russian Fleet 1966–1970

Fig. 23 Soviet Vessels Anchored off Destruction Island. Image #4693.

Cruise 81—My Seventeenth Cruise (1966)

I was asleep on the *John N. Cobb* when the change of tone of the low-speed diesel engine exhaust that had lured me to sleep woke me just before sunrise. The vessel had been running at a constant cruising speed throughout the night after leaving Seattle on September 19, 1966, and it had been uneventful. There was a light swell, and I'd been sleeping soundly. With the shift of the engine into neutral, I could feel the unexpected change, a definite slow rise as a swell lifted the ship, and then felt her gently fall as it passed. I got out of my bunk and headed up the steps to the pilothouse, sensing the coming daylight as I came onto the bridge. Pete, the skipper, glanced at me and said, "We have company." As I looked out the pilothouse windows, I saw a sight that stunned me.

There was no wind; the sea had a slow, glassy swell. To the east was the coast of Washington State with the Olympic Mountains in the background, dark in the predawn light, and everywhere else, there were ships, a giant fleet of nearly fifty Russian ships. As the sun rose, the red hammer and sickle on the stacks seemed to glow from the sun's reflection. A Russian city as close as three miles off the coast, unbelievable during the time of the Cold War, when we were lucky to see only a vessel or two off our coast as we worked or transited through the waters.

Looking around, I saw side trawlers (Fig. 24), as well as a few of the new 280-foot stern trawlers referred to as Russian class BMRT stern trawlers that caught fish and processed them in the factory below decks. One of the first Pushkin class was spotted. There were forty-two of this class of vessels built in West Germany for the Soviets between 1954 and 1956. This was the class of vessel Lee Alverson had seen in Europe in 1957 that he believed would change the distant water trawl fleets of the world by replacing the side trawler. The stern trawlers were a fresh concept to the distant water fleets of the world. The new design was large, high, and long, with a stern ramp where the trawl net would be hauled, dragging the catch of fish up behind it, similar to the whaling factory ships that pulled up the carcasses to be butchered. Once the fish were on the upper deck, they would be dumped through a hatch to the factory below decks, where they would be processed, frozen, and stored in the freezer holds. There were other ships, such as freezer ships and support vessels, that made up the Soviet distant water trawl fleet working off the Washington coast.

It was just so unbelievable that they were there at all! They targeted two fish—Pacific hake and POP—and were taking huge quantities of them. Pacific hake was located on the continental shelf, and they found

Fig. 24 Soviet Side Trawler off the Washington Coast, 1966. Image #4706.

the ones we were assigned to scout for. They also found the POP resource inhabited the upper edges of the continental slope.

The weekly movements of the fleet (Drawing 18) were first observed off Oregon in the second week of April 1966, with approximately twenty-nine vessels. They remained off Oregon for the rest of April and May. Moved to Washington in June and grew to 107 vessels by July 1966, where the fleet remained until the end of August. In September and October, they

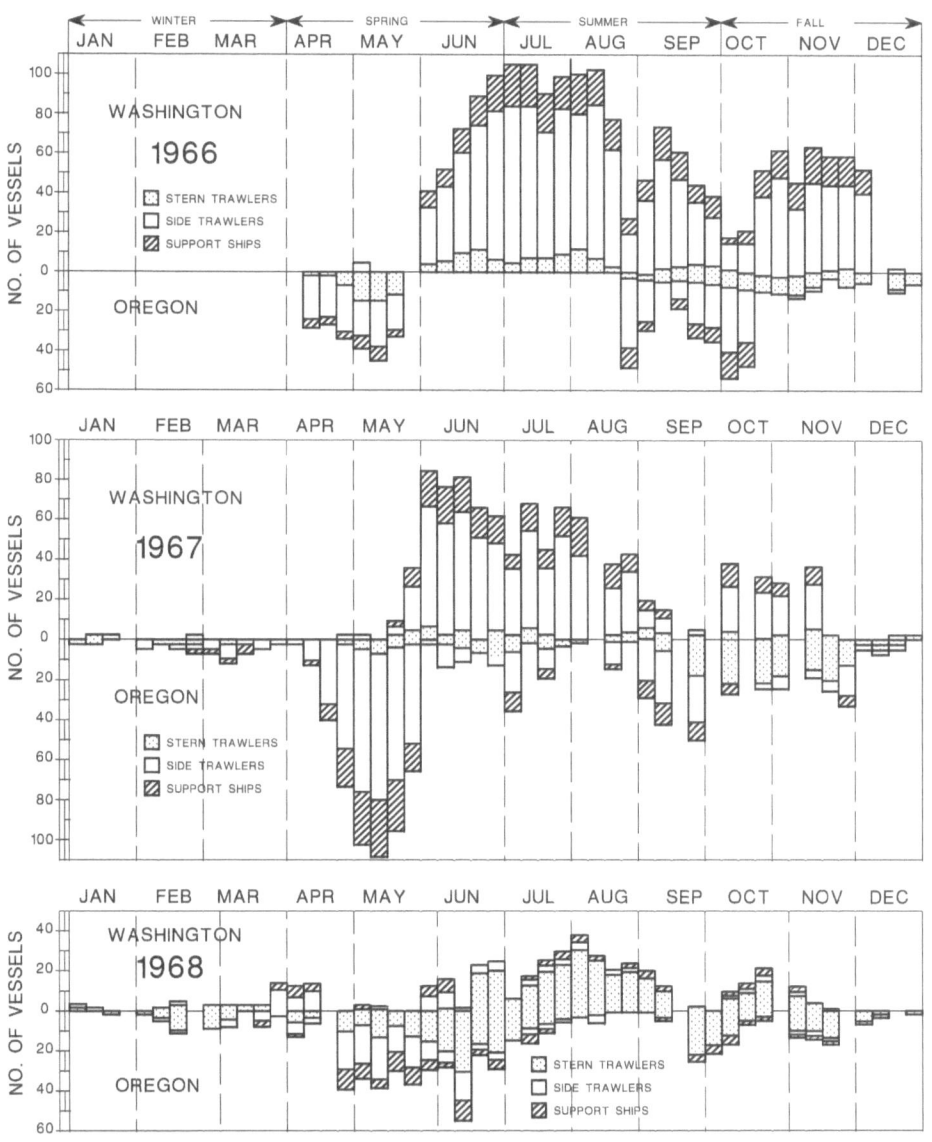

Drawing 18 Soviet Fleet Weekly Movements off the Washington and Oregon Coasts.

worked both Washington and Oregon waters. In November, they worked the Washington coast and had departed, except for a few stern trawlers, by December.[1] The first sighting from the *Cobb* occurred when the scientific staff reported them on Cruise 79, which occurred between July 11 and August 5, 1966. It reported that they encountered the Russian fleet with large freezer ships anchored between the 40- and 50-fathom contours between Destruction Island and Point Grenville (Fig. 23), with side trawlers working west of them. During the latter part of the cruise, the Russian fleet moved southward off Grays Harbor.

Before the fleet arrived, the Soviets had at least two exploratory vessels scouting the Washington and Oregon coasts in 1965, the side trawlers *Adler* and *Ogon*. Once the fleet arrived in 1966, the fishing vessels that actually caught fish could be divided into two groups: the side trawlers (classified by the Russian abbreviation SRT) and the stern trawler factory ships (classified as BMRT). The side trawlers needed to have support vessels where their catches could be offloaded into the freezer or refrigerated ships each day to preserve the catch for the trip back to Russia. The fleet needed additional support ships, such as fuel and water tankers (Fig. 25), supply ships, seagoing tugs, and personnel ships that looked like cruise ships (Fig. 26).

Fig. 25 Soviet Tanker. Image #4593.

Fig. 26 Soviet Passenger Ship with Side Trawler Alongside. Image #2802.

There were five American fishing vessels working the hake resource in 1966, all 70- to 80-foot long combination vessels and rigged with the midwater trawl and the electron cable system. They would deliver their catch to the Pacific Protein Corporation plant in Grays Harbor (Fig. 27). The American vessels were small compared to the Soviet ships and had a hard time competing with them. The Russians' massive fishing effort changed the schooling patterns of hake. Russian ships worked as a fleet, with a command ship as its heart, keeping track of the catches and location of each ship. When one came up with a good catch, others would be directed to the site, and once there, they would make a similar haul as the ship that found the fish. They would all go in the same direction and follow the same path. At the end, they would haul the net and retrieve the catch in the so-called "end zone." When the catch was aboard, each vessel would return to the start of the tow and repeat the process until the catch rates fell off. If one of the US fishing vessels got into the pattern, they were forced out of it just by the size and power of the Russian ships.

Fig. 27 Pacific Protein Corporation with *John N. Cobb* Tied Up Alongside. Image #5186.

By the time I went out on *Cobb* Cruise 81 between September 19 and October 14, 1966, the Russian fleet had been fishing hake and POP successfully and intensely off the Washington coast. During the time I was aboard the *Cobb,* there were approximately eighty vessels in the Russian fleet, and they worked both the Washington and Oregon coastal waters. There were more vessels than I had ever seen along our coast. Our assignment was to locate schools of hake to assist the five Pacific coast vessels that were fishing them. No large schools of hake were found during the cruise. Tracings observed on the sounder differed from those of previous cruises, not as compact nor as concentrated at any one particular depth. It reported that hake signs in the past were quite distinct and recognizable as hake, but during this cruise, echograms believed to be hake turned out to be large red jellyfish of the genus *Cyanea*. When caught in a trawl in large numbers, they caused a problem for the deck crew to clean the net. The species seemed to have been quite abundant that summer. I remember one part of this trip, even after all these years. We found a sign that appeared to be hake near the bottom on the sounder early one morning off the northern coast of Washington, so we set the gear and took a small catch of hake near the bottom. We continued to monitor the signs we believed were a school of hake. I radioed the American vessels that we had located a school and would monitor it.

We sounded the school. The signs came up off the bottom, and we stayed on them. Finally, a couple of the US fishing vessels arrived after steaming most of the day to the site. I told them where the signs were, and they set their gear, but when they hauled it back up, the catch was all red jellyfish. Somehow, we mixed up the schools and had sounded out a huge school of jellyfish. It was unbelievable they were as dense as they were. I'm sure those fishermen had unpleasant words about the CS on the *John N. Cobb*. The nets had to be washed out, and jellyfish stuck to the meshes like glue. Their tentacles, when contacting any exposed skin, stung. So, as the sun set, the *Cobb* slowly withdrew to the west into the sunset. Looking back at the fish boats with their nets in the air and the fire hoses washing off the remnants of jellyfish will remain in my memory forever. I can still hear Arnie singing his rhyme: "If you want it to go on the fritz, just call Hitz."

Nineteen sixty-six became the catalytic year with the appearance of the massive Soviet fleet, which took "our" fish in "our" waters right in front of us. POP was a new developing market of IQF fillets, for which the US military gave out large contracts. I believed that year was the start of changes that would affect the US fishing industry. The first thing that happened was that the three-mile limit was changed to a twelve-mile limit, which came into effect in 1967 to help the US hake plant. The Russian fleet stayed out of the twelve-mile zone in the following two years, and the American fleet grew to ten vessels in 1967 and could fish inside or outside the twelve-mile limit.[2] The US fleet still could not compete with the Russian fleet, which was one reason why the fish flour plant finally failed.

In 1967, the Soviets followed the same pattern as in 1966, with more effort off Oregon in April and May, with the peak of over one hundred ships occurring in the second week of May, then shifting to the Washington coast and peaking with over eighty ships in the first week of June (Drawing

18). Then, by the middle of July, the fleet declined to about seventy ships, primarily remaining off the Washington coast, though there were some on the Oregon coast. After the first week of August, the fleet size was reduced to about fifty ships. After that time, the fleet remained about the same size until the middle of November. There was an increase in the number of stern trawlers (Fig. 28) while there was a decrease in the other ships. By December, the fleet had disbanded, leaving a few ships in the waters off Washington and Oregon during December 1967 and January 1968.

Fig. 28 Modern Soviet Mayakovsky Class Stern Trawler. Image #4549.

I'm not sure if the fish flour plant closed in 1968. However, I kept track of the Soviet vessels during 1968 and added it to Drawing 18.[3] The fleet size was reduced by half to about fifty ships in June. The pattern was somewhat similar to the previous years, starting off Oregon with a few stern trawlers. By the end of April, a fleet of about forty vessels dominated by side trawlers remained off the Oregon coast until the second week of June, when the stern trawlers dominated the fleet, which reached its maximum size of about sixty vessels. After that time, the side trawlers disappeared with their support vessels, and by the end of June, there were about forty stern trawlers and a few side trawlers left.

The Pacific Fisheries Management Council in 1987 presented a table in a publication that listed the total catch of POP in metric tons of fish taken by year in the area off the coasts of Oregon, Washington, and the east side of Vancouver Island south of Cape Scott.[4] I have converted metric tons to pounds landed by year and presented them in a graph showing the results (Drawing 19). When the Soviet Fleet arrived in 1966, there was a rapid increase in the catch to 61,000,000 lb. of POP, which caused a lot of concern and is why I believe 1966 became the catalytic year of change. The catch peaked in 1967 at 82,000,000 lb. Then it plummeted to 44,000,000 lb. in 1968 and to 13,000,000 lb. in 1969, collapsing the POP fisheries.

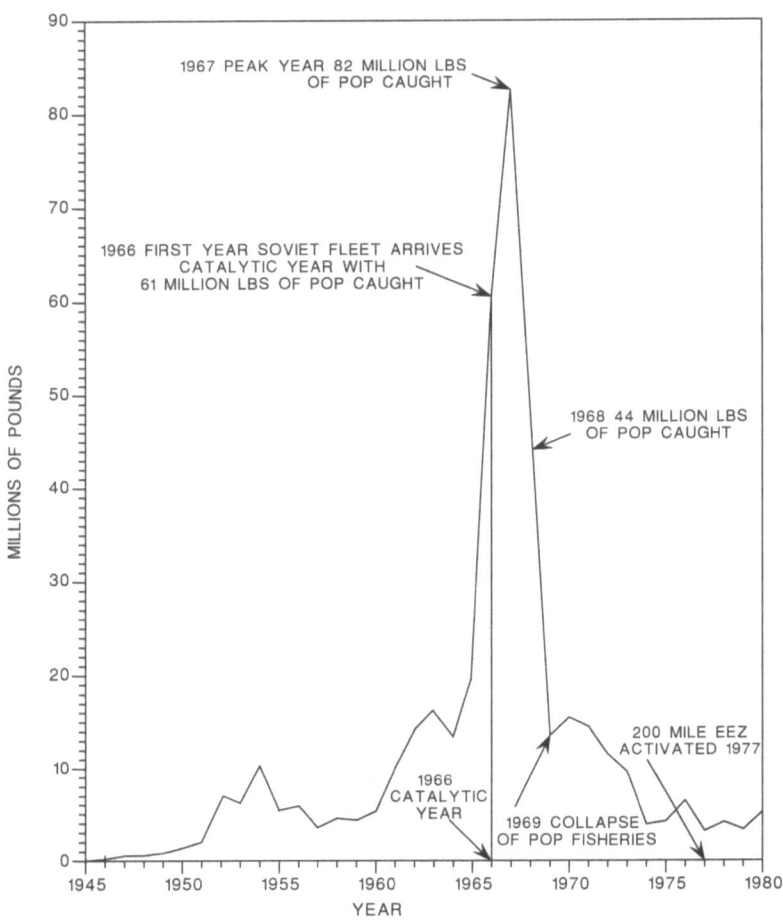

Drawing 19 POP Landings from Waters off Oregon, Washington, and Vancouver Island.

I'm not sure what happened to the Soviet fleet after 1968. I know the Russians needed the fish because they worked out a system with the US regarding hake. The smaller US combination fishing vessels would catch the hake and transfer the catch by disconnecting the cod end of the trawl with the catch and floating it over to the 278-foot stern trawler factory ships standing by. The Soviets would haul the cod end up the ramp, open the fish hatch, and dump the catch below decks to be processed in the factory below. This fishery was referred to as a joint venture, and the Soviets continued fishing hake but not POP. Hake are fast growing and the majority of them caught in the nets are five to seven years old.[5] They leave the area in the winter. Since they are processed quickly, they are not mushy so are now listed as food fish and are processed into fish sticks.

The common name "hake" has been changed to "Pacific whiting" and is a prime resource sought after and harvested by the US trawl fleet today. If they are managed correctly, the Pacific whiting fisheries should last.

There has been a debate on whether POP were slow growing or not. There are two methods of determining age for POP. The first was by reading the growth rings on an individual's scale, like counting the growth rings on a tree. The second was the use of otoliths or ear stones. It was believed they would die at about twenty years old. The Soviets and Japanese continued using scales. The US scientists started counting the rings on the otoliths, which lie in the head and are used to keep the fish's balance.[6] They found that the fish stopped growing; the energy was apparently put into reproduction instead of growth. Once the growth stopped, the growth rings did not show on the scales, but they did show on the otoliths. They found that POP were slow growing and continued living until ninety years old. Removal of the large fish that took years to grow left a void of spawning stock, and the question is whether the POP fisheries will ever come back.

Catalog of the Russian Fleet

I became even more interested in the Soviet fleet after hearing what the American trawl fleet said about the US government. They told us that when they asked for information from the Coast Guard about what the fleet was doing along the Washington and Oregon coasts and what types of ships they were, they would say, "We can't tell you because it's classified information." This angered the American fishermen. Since I wasn't cleared to see our government-secured information, I went to the University of Washington Library and looked at Russian publications, which had drawings of their vessels. I had already done this with the *Adler*. Many of those articles had already been translated into English by the federal government and were available to the public.

Using this information and drafting the outboard profiles of the Russian vessels, I developed the "Catalogue of the Soviet Fishing Fleet," published in the *National Fisherman* yearbook issue in 1968.[7] Reprints were made, so we could go down to the American fishing boats and give them a copy, asking them to identify the Russian vessels. They could then let us know what the Soviets were fishing and where. This satisfied them, and we received much information on the Soviet fleet's activities. We also gave them to the Coast Guard, who could then distribute them to interested parties, as they were not classified.

Fig. 29 Washington State Piper Aztec Aircraft
(left to right, Gene Dinitoa and Brad Pattie). Image #4586.

In 1967, I was invited to fly over the Russian fleet with the Washington State Department of Fish and Wildlife because I was developing a catalog of the Russian fleet. It's now been fifty years since I stepped onto the plane at the airport in Olympia, the capital of Washington State. It was a twin-engine Piper Aztec that could seat six individuals. I sat on the left-hand side behind the pilot and the co-pilot, both Washington State Patrol officers. The plane was the governor's. Brad Pattie and Gene Dinitoa worked for the State Department of Fish and Wildlife and were also aboard. Brad sat in the last seat so he could move to one side or the other, depending on which side of the ship they passed. We taxied out to the end of the runway and stopped. We checked that our seat belts were fastened and had on the life vest, so-called May West. Then we were off with the engines at full throttle, gaining speed down the runway. Once airborne, the wheels came up. The pilot then set a course for the ocean.

I had the opportunity to interview Brad Pattie on October 23, 2012, after many years of not seeing each other. He flew on seven of the eight flights over the fleet and supplied me with information about them. He said that the commercial trawl fleet wanted to know what the Russians were doing along our coasts, how many ships there were, and what they were catching.

On this flight, he recorded each individual ship that we passed by name and hull number and took pictures of each vessel as the plane flew by. Meanwhile, the pilots recorded two different loran readings off a machine in the cockpit as they passed by, which were converted into longitude and latitude later and went into the report. We began the flight in Olympia, flying at one or two thousand feet, coming out near Grays Harbor, Washington, and flying out in a diagonal direction from the coast, looking for ships at this height until we reached twenty miles off the coast. Then we changed course, heading back in toward the coast. Whenever a ship or vessel was sighted, we would rapidly descend to just above the ocean's surface and fly by the side of the ship. Brad would note it in his notebook. If it was a new or unknown ship, we would return and take pictures of it, making sure we got the name and numbers off the hull and house.

It took about two hours to cover the coast of Washington out to twenty miles, and when that was done, they landed at a coastal airfield and took a break. Afterward, they would fly out forty miles and cover the area between twenty and forty miles along the coast, including the 100-fathom contour where the shelf became the slope. When they flew over vessels, they'd check the catch if they had one. If the fish seen in the net or on the ship were colored red, they were probably POP, and if they were silver, they were probably hake.

When we first spotted a vessel, we banked and rapidly dropped, losing attitude until we were just above the water's surface. The plane cruised at about 180 knots. As we neared the vessel, the pilot lowered the flaps, slowing down to about 110 knots. As we passed, Brad read the numbers and name and recorded them in his notes. I took it all in, an experience I will never forget. It was exciting! We did it time after time. At least thirty to forty vessels were counted that day. The pilots were extremely skilled. Brad said that on one of their flights, they went alongside a large support vessel, and as they passed, he looked up at the bridge. They were that low.

After the morning flight covering the coastal area out to twenty miles, we took a break at Copalis Airport in Grays Harbor. It was a relief to get out of the plane and walk around. Later, we took off and flew out to forty miles off the coast and finished off the coastal area between twenty and forty miles. We saw large stern trawlers hauling their nets up their ramps, likely POP. Every time we banked, let down, and buzzed a ship, it was a thrill. It must have been similar to what the pilots did during World War II.

We returned, flying over Astoria, Oregon, and could look down and see the new bridge, where the *Cobb* was tied up at the port docks when we worked the AEC stations off the Columbia River. Then we flew low along the coastline above the surf line until we gained altitude and headed east for Olympia Regional Airport, completing a wonderful day, a once-in-a-lifetime experience.

Fig. 30 Soviet Stern Trawler off the Washington Coast. Image #4630.

After flying over the fleet and observing Soviet side trawlers and stern trawlers fishing off the Washington coast, I started wondering what made these large stern trawls so unique that they replaced the side trawlers that for years had proven to be wonderful sea boats that endured the winter storms in the North Atlantic. I had the fortunate opportunity to accompany a modern US 295-foot stern trawler, the *Seafreeze Pacific*, as well as the *G. B. Reed*, a side trawer.

Chapter 11: Endnotes

1. Charles R. Hitz, "Operation of the Soviet Trawl Fleet off the Washington and Oregon Coasts during 1966 and 1967," *Pacific Hake*, Circular 332 (1970), pp. 53–75. MS #194.

2. Martin O. Nelson, "Pacific Hake Fishery in Washington and Oregon Coastal Waters," *Pacific Hake*, Circular 332 (1970), pp. 43–52. MS #202.

3. Charles R. Hitz, "Soviet Hake Fleet Keeps Up Pressure," *National Fisherman* (April 1969). MS #208.

4. Daniel H. Ito, Daniel K. Kimura, and Mark E. Wilkins, "Status and Future Prospects for the Pacific Ocean Perch Resource in Waters off Washington and Oregon as Assessed in 1986," NOAA Technical Memorandum NMFS F/NWC, 113 (April 1987).

5. Martin O. Nelson and Herbert A. Larkins, "Distribution and Biology of Pacific Hake: A Synopsis," *Pacific Hake, Fish and Wildlife Circular* 332 (1970), pp. 23–33. Age and growth on page 30.

6. Donald Gunderson, *The Rockfish's Warning* (Seattle: University Book Store Press, 2011), pp. 120–122. Pacific Rockfish: Abundant, Long-Lived, and Rapidly Depleted.

7. Charles R. Hitz, "Catalogue of the Soviet Fishing Fleet," *National Fisherman Yearbook Issue* 48, no. 13 (March 31, 1968), pp. 9–19, 21–24. MS #172.

CHAPTER TWELVE

Seafreeze Pacific & G. B. Reed

Fig. 31 R/V *Seafreeze Pacific*. Image #5481.

Al Pruter asked me to accompany him to Bellingham and look at a new 295-foot stern trawler that had just arrived from the East Coast on October 28, 1969. I jumped at the chance. However, he had something more on his mind. The vessel was moored at the Bellingham cold storage pier (Fig. 31). As we approached her, we saw the name *Seafreeze Pacific*. This ship, a sister ship to *Seafreeze Atlantic*,[1] was the second factory stern trawler built for the United States fishing industry under the 1964 Fishing Fleet Improvement Act. The *Seafreeze Atlantic* would fish ocean perch, *Sebastes marinus*, on the East Coast, whereas the *Seafreeze Pacific* would fish POP along the Pacific coast. They'd operate among the foreign trawlers using similar fishing gear, often fishing on the same grounds. They were built because, when the Soviets' distant water fleet began to trawl the Gulf of Alaska for POP, the industry was irritated and wanted something done.

The decision was made to build two ships, one for the Atlantic and the other for the Pacific, to compete with the Soviet vessels.

I was anxious to go through the ship and try to understand why this type replaced the distant water side trawlers in the world fleets. We met the skipper, Erling Jacobsen, and the plant manager, John Schmiedtke, and talked about their first cruise and when they planned to sail. The trip would start sometime in November, be at sea during Thanksgiving, and end before Christmas. This is the time when fishermen in this region begin shutting down for the oncoming winter weather. After looking over the deck and fish-processing area of the ship, we looked at the stern ramp, which was the item that made this ship so important. It was where the cod end of the net with the catch would be pulled up the ramp onto the fish deck (Fig. 32).

Fig. 32 R/V *Seafreeze Pacific* — The Stern View of the Ramp. Image #5473.

On the way back to Seattle, Al, in his diplomatic way, said they would like me to be the observer on this first trip as a guest of the company that was running the vessel. He said I had knowledge of the trawl fisheries and knew about the operation of the Russian fleet and their ships. It didn't take much time for me to agree to go. I departed with the ship on November 14, 1969, from Bellingham on her first cruise, which ended on December 14, 1969.

Stern Trawler — *Seafreeze Pacific*

I had the run of the ship and could go anywhere I wanted to. I had days to explore the ship. I was interested in the fish deck, where the nets were set and retrieved. I was also interested in where the fish went when caught and how the POP were separated and processed into IQF fillets below decks in the factory. The only place I didn't inspect was the engine room. She was a diesel-electric ship. The engine room produces electricity, which is used to drive an electric motor which turns the ship's propeller to drive the ship.

The first thing I had to do was to understand the fish plant below deck (Drawing 20), where I needed to be when the hatch near the stern was open and the entire catch was dumped into the fish bins. The catch would have to be sorted by the crew. POP, the targeted species, would be placed on the belt that ran along the face of the fish bins, which took them to the washer and then to a holding bin. The larger fish, such as lingcod, true cod, sablefish, and rockfish other than POP, were set aside in the checkers. Trash fish, such as dogfish and turbot, were placed on the lower belt that took them to the trash chute, where they were dumped overboard or sent to the fish meal plant in the after part of the ship.

With time, a procedure was set up. I would observe whether the fish bins below deck were clear of the previous catch when the next haul was retrieved. Above, on the fish deck, I would observe the cod end as it was pulled up the stern ramp (Fig. 33). Then I'd watch the deck gates close. The cod end was opened after it was pulled forward of the hatch. The hatch was then opened, and the catch spilled into the bins as the opposite end of the net was lifted up by the line hanging from the overhead stanchion. I would make an estimate of the total weight of the entire catch by how full each of the four bins were, whether to the top, halfway, a quarter way, or an eighth of the way (Fig. 34). I would also get an idea of what species were in the catch as it spilled out of the bag and into the hatch.

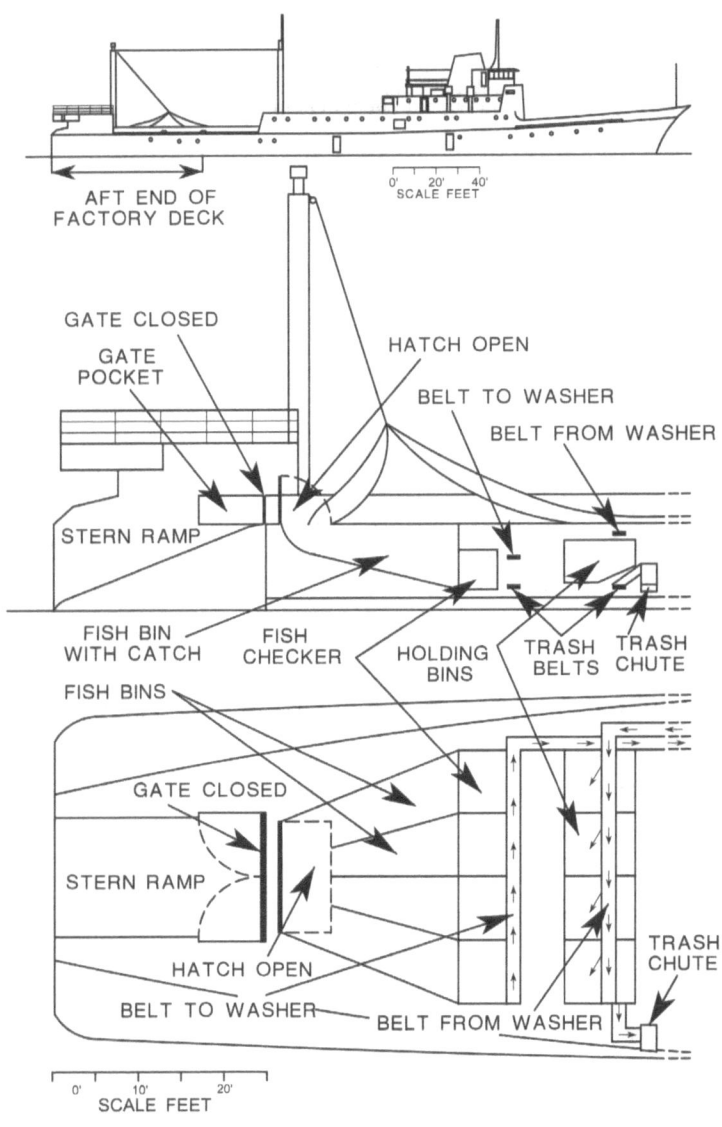

Drawing 20 *Seafreeze Pacific*—Aft End of the Factory Deck.

Once I was below deck, I had another chance to make my estimate. I saw how much of the bin was filled and would make an estimate of the composition of the catch, such as the percentage of POP and other species. I checked the bins for halibut and salmon. There was considerable concern about the amount of salmon and halibut the foreign trawlers were taking. Observing the *Seafreeze Pacific* catch, we could assume that similar catches were made by the foreign trawlers.

Fig. 33 R/V *Seafreeze Pacific*—Cod End of the Net Pulled Up the Ramp. Image #5555.

Fig. 34 R/V *Seafreeze Pacific*—Fish Hatch Open to Fish Bin. Image #5559.

Early detection of halibut was important. They slid to the foot of the fish bin near the bin boards, where they could be found rather quickly. About 90 percent of them could be found right away. If they were returned to the water fairly quickly after detection, their chances for survival were good. I looked for any tagged halibut, measured them for total length, and then dumped them through the trash chute, hopefully, to live longer in the ocean. The rest of the halibut were measured and dumped into the ocean through the chute, and some would survive.

Salmon, on the other hand, were mixed in with the catch and were more difficult to find. A total of eighty-three were recorded caught during this cruise, all Chinooks (*Oncorhynchus tshawytscha*), with an average size of twenty-four inches. Most were taken in two different hauls along the Washington coast. They were a beautiful prime fish but in poor condition after being taken in the trawl. Once I found them, I recorded their total length and then released them back to the sea through the trash chute. I observed what I called "salmon fever"—when a crew member spotted one, he would pick it up and hide it in the freezer, so he could secretively take it home. All I wanted to know was what species it was and its total length. It was important to know because it would give us an idea of what salmon the Soviet vessels were accused of taking.

Once the factory crew began to sort the catch, the POP were placed on a belt, which took them to the washer and onto the holding bins. I would take a number off the belt at a time, measure them for total length, determine whether they were male or female, and place them back on the belt to continue on their way. I'd record some of that information in a fishing log kept on the bridge, one that Lee Alverson had instigated for the State of Washington to use when he worked for them. While I was aboard, I kept it up to date. The total estimated weight of the catch by species was recorded in the log by haul number, along with date, loran readings, depth, and time of the set and end of the haul, which were obtained from the bridge.

I had the opportunity to observe the fishing technique of a newly built stern trawler the 296-foot *Seafreeze Pacific* that was built in America and was delivered in May 1969 (Drawing 21). It was designed to compete with the Soviet's 277-foot *Pushkin* class vessel. Lee Alverson had the opportunity to visit one of them in Germany in 1957. The *Pushkin* was the first vessel built in a class of forty two vessels by West Germany between 1954 and 1956 for the Soviets. They were built on the new fishing technique using a unique stern ramp to land the fish that it caught. The following is what I observed about how the fishing technique worked while I was aboard the *Seafreeze Pacific*.

The drawing is divided into four parts. Item 1 is the top view of the fish deck setting the trawl. Items 2 and 3 show the joining of the trawl doors

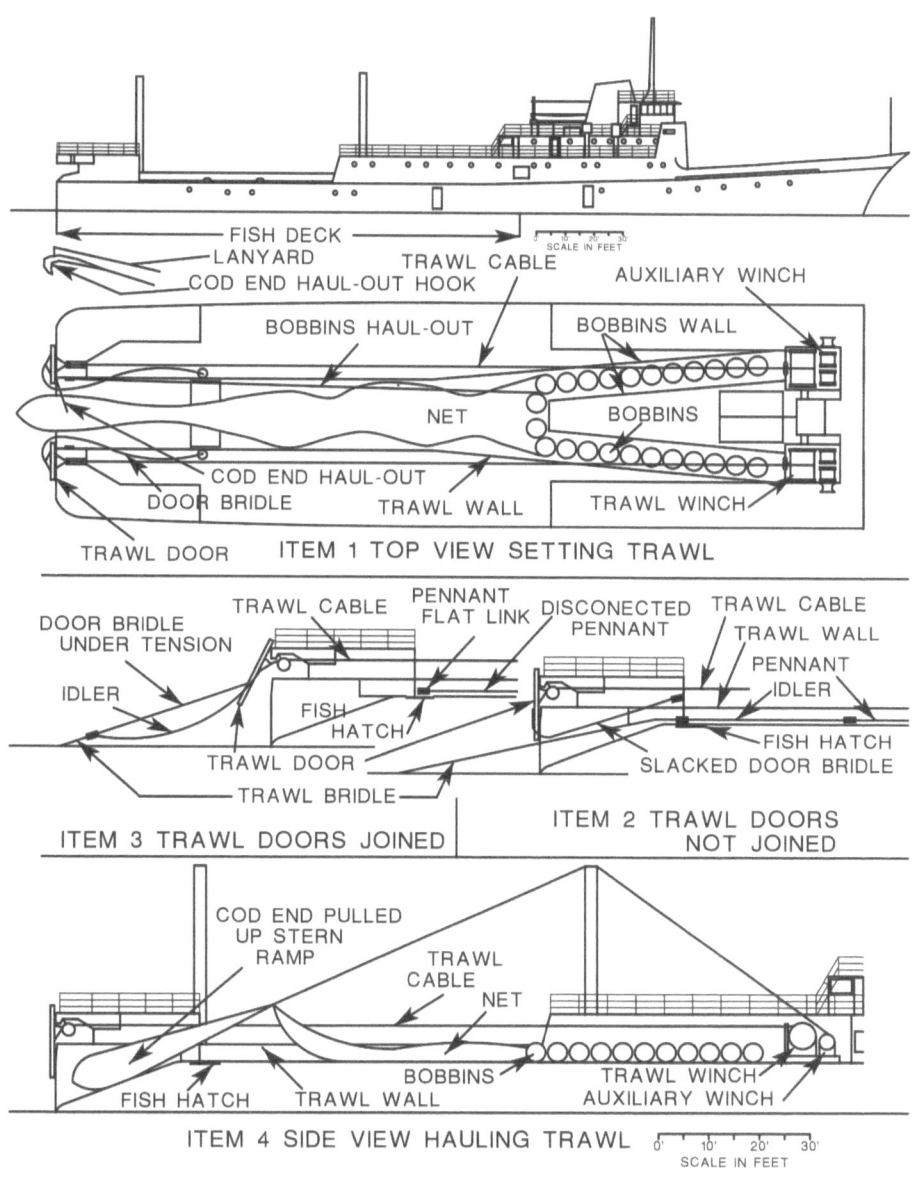

Drawing 21 R/V *Seafreeze Pacific*—Fish Deck.

to the net, and Item 4 shows the hauling of the cod end. Item 1: The net is pulled along the deck by hooking the cod end out-haul cable into the webbing. It is pulled onto the stern ramp and into the water. The hook is freed by using the lanyard attached to it. The vessel slowly moves forward on a straight line as the rest of the net is pulled out of the stern ramp by the drag of the net in the water. Once the end of the net reaches the attachment of the bobbins, the net is stopped. The weight of the bobbins is too great for the drag of the net to move them. The bobbin out-haul cable is hooked onto the bobbins and is pulled along the deck into the stern ramp. There is enough drag of the net in the water to pull the rollers out of the ramp once the hook is released by a lanyard. The ship continues moving forward with the bobbins and net being pulled off the vessel along with the trawl bridles—one on the starboard side and the other on the port that are each attached to the bobbins and the wings of the net.

The other end of each of the trawl bridles is attached to the idler and then a pennant that is attached to the drum of each of the auxiliary winches. The breaks on the auxiliary winches are applied to stop the trawl bridal connection with the idler going past the fish hatch. The flat link that joins the trawl bridle and idler is under tension from the pull of the net and bobbins (Item 2). At this point, tension of the net is transferred to the trawl doors by connecting the slacked door bridle that contains a G-hook that is attached to the after end of each trawl door bridle. It is clipped into the flat link that connects to the trawl bridle and idler. The auxiliary winches are released until the strain of the trawl is taken up by the doors (Item 3). The slack idler is disconnected from the pennant and is attached to the door. The trawl doors are then let out evenly to the desired depth, and the deck gates are closed.

Item 4: Once the haul is completed, the trawl cables are reeled in on each of the trawl winches until the doors reach the stern stanchions. The slack idler is taken from the door and attached to the disconnected pennant lying on the deck. Each auxiliary winch is engaged winding the pennant followed by the idler, and then trawl bridle until the bobbins appear on deck. Each wing of the net attached to the bobbins are pulled the rest of the way along the deck by attaching a bobbin haul in. The net and bobbins stay within the bobbin walls that forms the container just aft of the trawl winches. The rest of the net follows behind the bobbins, staying inside the trawl walls. Once the bobbins are in their container and the intermediate part of the net is pulled up on deck, a strap is placed around the rest of the net and the cod end is pulled up the ramp until it rests just forward of the fish hatch. The deck gates are closed, and the fish hatches opened. The cod end is opened and then lifted, and the catch dumped into the fish bins below deck.

The impression I got from observing this operation was the size of all the equipment involved to move the bobbins. The modification of the *Cobb* for the AEC work in 1963 was similar to the *Seafreeze Pacific*. Both vessels had the trawl cable from the trawl winches connected to the trawl doors. The net reel on the stern or the *Cobb* was used to store the trawl bridals like the auxiliary winches did on the *Seafreeze Pacific*. On the *Cobb*, the reel also rolled up the net. The *Seafreeze Pacific* had space to store the bobbins and the equipment to move them.

Fig. 35 R/V *Seafreeze Pacific*—Bobbins Ready to be Pulled Up Around the Container. Image #5524.

SIDE TRAWLER—G. B. REED: The year before, in April 1968, I went out on the *G. B. Reed*, a 177-foot side trawler and a rarity on the Pacific coast. CS Jergen Westrheim invited me to join him on one of his rockfish cruises because of my interest in the early life history of the rockfish and to test my new field key to aid in their identification. I believe she was the only side trawler operating along the Pacific Coast at the time, before the Soviet fleet arrived in 1965. I had the opportunity to observe the actual fishing procedure of a side trawler.

The *G. B. Reed* (Fig. 36) was designed on the basic fishing practices of European side trawlers, which had proven themselves since 1885.[2] The German side trawler was a good example,[3] as well as the British deep-sea trawlers.[4] The fishing technique was describe by Boris O. Knake in 1958 in "Operation of North Atlantic Type Otter Trawl Gear."[5] He had the talent to illustrate the entire process. They were an excellent sea boat. The house and engine room were in the after part of the ship, and the main landing

Fig. 36 Canadian R/V *G. B. Reed*. Image #5036.

area of the fish catch was forward of the house, where the skipper could watch over most of the operation. The low freeboard in the ship's forepart helped to control the heavy lift of the fish from the sea by lifting the cod end and sliding it up against the hull, then over the rail, where the catch was dumped into the checkers for sorting. The high forepeak of the bow protected the crew from heavy seas while working on deck.

Bobbins were heavy, somewhat rigid and inflexible. They could be stored just inside the starboard rail between the fore and aft stanchions. When in use, the bobbins would be attached to the trawl net's foot rope. The net would be placed in the water, followed by the bobbins that were lifted with overhead tackles and placed in the sea. When the haul was completed and the catch brought aboard, the bobbins could be lifted from the water and placed on the deck, just inside the rail. They were used so the net would pass over rough bottoms that snagged and hung up trawls that tore up the nets.

When I was on the *G. B. Reed*, I got to see the method they used to set and retrieve the trawl (Drawing 22). Once the drag was decided on, the vessel was put into a circle or a large arc as they played out the net after the trawl doors were connected to the trawl cable (Fig. 37), as they did on the *Cobb*. The same procedure was used when I went out on my first trip on the *Cobb* in 1960, the difference being that the *Cobb* towed the net from the stern, not the side.

Once the doors were in the water on the side trawler and the net was spreading, the vessel was then lined up for the drag. The amount of cable for the depth of the drag was let out, then the winch brakes were set. The

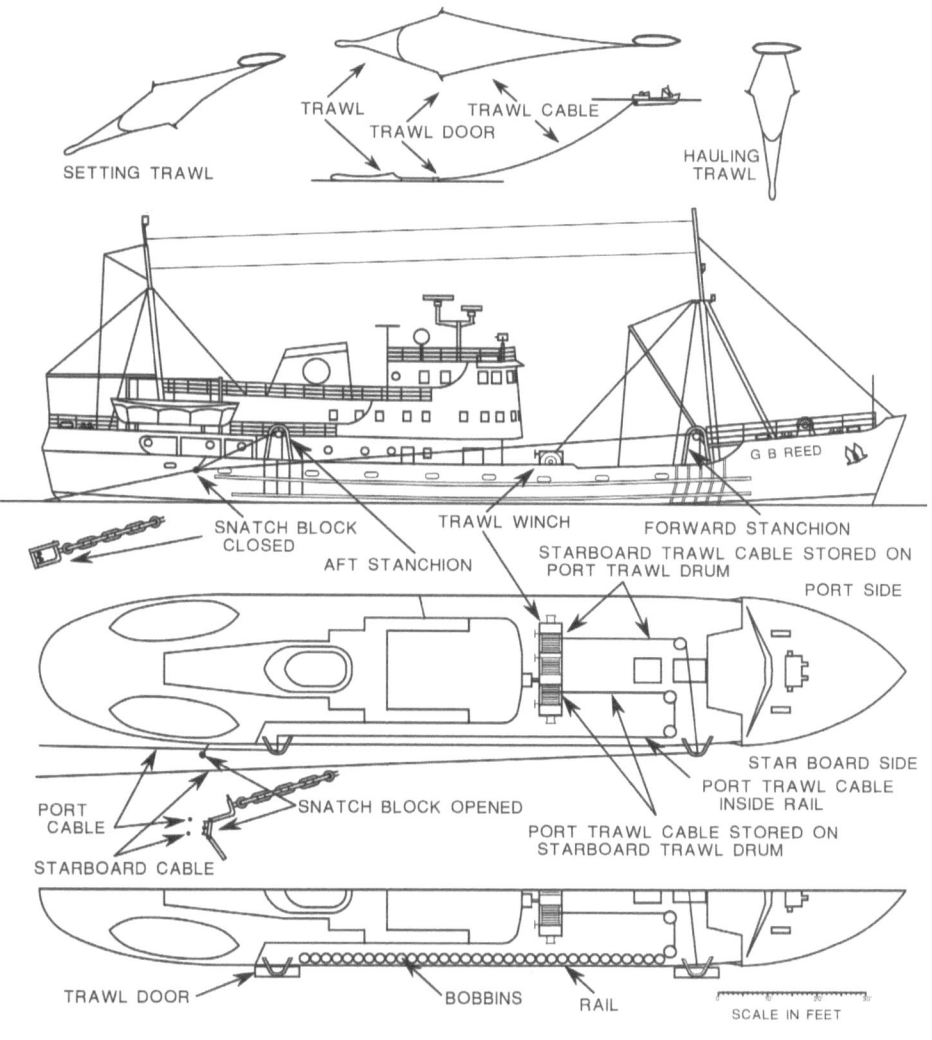

Drawing 22 Canadian R/V *G. B. Reed*—Fishing Arrangement.

crew snagged the starboard cable leading from the forward stanchion with a device that could be used to pull the two cables together into a snatch block attached to a short piece of chain attached to the vessel just aft of the aft stanchion (Fig. 38). When closed, the snatch block joined the two cables to a single towing point where the trawl net was pulled, hence the term "side trawler" (Fig. 39). The two cables remained together during the remainder of the haul.

Connecting and disconnecting the trawl doors from the trawl cable and placing the cable in the snatch block and releasing it was a dangerous procedure because of the forces involved, especially in bad weather.

Fig. 37 Canadian R/V *G. B. Reed*—Connecting the Trawl Door to the Main Trawl Cable. Image #4920.

Fig. 38 Canadian R/V *G. B. Reed*—The Outboard Cable Pulled In. Image #4910.

Fig. 39 Canadian R/V *G. B. Reed*—The Two Cables Placed in the Snatch Block. Image #4914.

When the haul was finished and the skipper gave the command to retrieve it, the crew pulled the pin holding the two cables together, and the vessel began to haul in the cable. Once the net was off the bottom, they turned to starboard, so she was broadside to the net as it came in. The doors arrived at the stanchion and were disconnected from the main trawl cables, and the bridles were wound onto both the main winches. Once the wings arrived at the stanchion, a lazy line that was attached to the head rope could be reached. The other end was attached to the cod end and, by pulling the line in, brought the cod end to the ship. Then it was lifted from the water by an overhead tackle, sliding the bag up along the side of the hull and over the rail, dumping the catch into the checker (Fig. 40).

Fig. 40 Canadian R/V *G. B. Reed*—The Cod End Being Lifted Aboard. Image #4999.

The following lists eight items I believe caused the Soviets to replace the side trawlers with stern trawlers. To help understand the description of the items refer to Drawings 21 (stern trawler) and 22 (side trawler).

One: stern trawlers could hold a straight course when setting the trawl and doors. The side trawler, on the other hand, had to maneuver in an arc to get the net and doors out so the net could fish properly.

Two: bobbins that were used some of the time were stored inside the rail on the side trawler, while on the stern trawler, they were stored on the fore part of the fish deck in a bobbin container. When the bobbins were used on the side trawler, they had be attached to the trawl foot rope and then lifted up and over the rail and placed in the water on the side trawler. Whereas on the stern trawler, the foot rope of the trawl was attached to the bobbins. Once the net was in the water, it acted as a sea anchor as the vessel moved slowly forward. The rest of the net was pulled off the vessel until the bobbins that were too heavy stopped the net. The bobbins then had to be pulled along the deck with a haul-out cable to the stern ramp, and then the trawl already in the water would pull the bobbins off the rest of the way into the sea.

Three: the trawl cables on the side trawler had to be the same length at each stanchion, so the trawl could be set evenly. The starboard side trawl drum had to let out more cable than the port side drum. Whereas on the stern trawler, they were of equal length.

Four: once the trawl gear was set on the side trawler, the two cables were connected near the stern by the crew in a snatch block. On the stern trawler, cables were the same length, so the gear could easily be set from the stern, which improved the crew's safety.

Five: on the side trawler, the main trawl cable was not attached to the trawl doors. The doors had to be attached and deattached from the trawl cable each time a haul was made. The net was attached to the main trawl cable via the trawl bridle, a jam link, an idler. All, with the exception of the net, were stored on the trawl winches. On the stern trawler, the main trawl cables were attached to the trawl doors, so the net bridles and idlers were stored on small separate winches via a pennant, making it much easier for the stern trawler to connect the trawl doors with the net and safer for the crew.

Six: when hauling the trawl, the side trawler pulled the trawl doors to her side along with the net, forcing the vessel to be broadside to the trawl and the wind. On the stern trawler, the net was lined up with the stern of the ship, and the trawl could easily be pulled up the stern ramp.

Seven: the catch could be dragged up the stern ramp and dumped through a hatch to the factory below on a stern trawler. On the side trawler, it had to be lifted over the side rail and dumped into the checker, where the catch was sorted on deck, and the commercial fish were placed in the fish hold, where they were preserved for future processing.

Eight: the factory below decks on the stern trawler was another factor that made them more desirable, reducing the support ships needed, as the Soviet vessels demonstrated in 1968. The stern trawler was easier to operate, safer for the deck crew, and processed the catch on the ship.

Seafreeze Pacific — Trips (1969-1970)

There were five trips made by the *Seafreeze Pacific*.[6] Trip 1 was the shakedown cruise. Two observers were put on the vessel, both from the US Fish and Wildlife Service Montlake Lab. The biological observer who supplied a report on the catch, fishing gear, and vessel was from the exploratory base. The other observer was from the Technological Unit, who reported on the processing and freezing phase of the catch on the factory below deck. On Trip 2, the International Pacific Halibut Commission supplied a biological observer to obtain catch data and tag incidentally caught halibut, supplying the second report. The Fish and Wildlife Service supplied a fishery biologist from Montlake Lab's Biological Unit for Trip 3, and he completed the third report. The US Fish and Wildlife Service also supplied an observer from their Juneau, AK, exploratory base for Trip 4. He was on the vessel for only the first half of the trip and supplied the report for the time he was aboard. There was no observer aboard the vessel during the last half of Trip 4, nor for the entirety of Trip 5. The skipper kept the Washington State Fisheries Log when no observer was aboard. The total crew size was fifty-six, divided into deck, engineer, and fish factory departments. There were officers, the captain and two mates, the chief engineer and his two assistants, and finally, the plant boss. The fishermen, divided into two groups of seven people, worked six hours on and six off, seven days a week while at sea.

The trawl fishermen were Portuguese and very knowledgeable. To get enough plant workers, the Bellingham jail was emptied, and the prisoners were sent to the ship to complete their sentences at sea. While we were eating the evening meal in the officers' mess at sea one night, the skipper suddenly got up and went into the crew's mess, where there was some kind of uproar. I believe he said something like the following: "If this doesn't stop, the crew members causing this will go into the brig. You are at sea, and the captain of this ship has dictatorial power and will use it. The brig is in the bow of the ship, the most uncomfortable place, and I can lock you up and place you on bread and water until we reach port, where you will be led off in handcuffs and turned over to the civilian authorities." There were no more problems after that day.

During the first part of the first cruise, the vessel proceeded up the outside of Vancouver Island to Queen Charlotte Sound, making a couple of hauls on the way up just to get the crew familiar with the equipment. When they arrived in Queen Charlotte Sound, they made several tows, which they compared to the trawl hauls made by three American trawlers working there: *Soupfin, Windjammer,* and *Don Edwards.* They outfished the *Seafreeze Pacific* two to one, but when modifications were made to the fishing gear of the *Seafreeze Pacific*, with larger doors and longer bridles, the gear seemed to work better.

The target species was POP, and the depth they looked for tows during this cruise was from 100 to 180 fathoms along the top of the continental slope. The skipper would inspect an area he knew had produced POP in the past by sounding it with an echo sounder. If it showed no fish signs, he would move on. If the sounder showed signs of fish, he would make a trial tow to determine what species the sounder showed. If it was a good catch of POP, he would continue making the same haul until the catch rate dropped off.

When I returned from the first trip, I was given the responsibility of keeping track of the reports from the remaining four trips, the last ending on December 23, 1970. Both ships were retired on that date. Both were later sold and, as far as I know, are working in the Alaskan Pollock Fishery. The *Seafreeze Atlantic* was renamed *Arctic Trawler* and then *Seafreeze Alaska.* The *Seafreeze Pacific* was renamed *Royal Sea* and then *Katie Ann.*

Table 7 shows the total catch of 4,758,365 lb. made in the five trips the *Seafreeze Pacific* made. The amount of POP taken during the trips was 3,162,580 lb. and made up 66.5 percent of the total catch. The percentage of other rockfish taken, excluding POP, was 3.5 percent. Trip 2 was cut short because of modifications that had to be made in the factory.

Table 7 *Seafreeze Pacific's* Reports.

Trip	Date Start End	Total Catch (lb).	Total POP (lb)	%	Total Rockfish (lb)	%
1	11/14/69–12/14/69	566,550	309,100	54.6	17,850	3.15
2	3/27/70–4/7/70	150,300	115,475	76.8	5,350	3.56
3	5/28/70–7/13/70	879,615	624,005	70.9	63,496	7.22
4	7/27/70–10/14/70	2,249,900	1,410,800	62.7	73, 400	3.26
5	10/25/70–12/27/70	912,000	703,200	77.1	8,000	.88
	Total	4,758,365	3,162,580	66.5	168,096	3.53

The biological observer Herbert H. Shippen who was aboard the vessel during Trip 3 kept detailed notes and records of the catches of all the fish taken on his trip. The percentage of rockfish, excluding POP, taken in the total catch was 7.22 percent, the highest compared to the other trips. I assume it was because the observer put a lot of effort into determining individual species taken in each haul. The other observers' percentage was 3.15 to 3.56, with less attention paid to the makeup of the individual species of rockfish taken in the catches. Trip 5 had no observer aboard, and the skipper didn't have the time or personnel to get the other rockfish count, which shows in the table as 0.88 percent of the rockfish catch.

The observer's records for Trip 3 were well organized. He recorded a total of 127 hauls made in five different areas during the trip: Area A (Washington coast) had 20 hauls, Area B (Vancouver Island) had 7 hauls, Area C (Queen Charlotte Sound) had 24 hauls, Area D (Southeast Alaska) had 5 hauls, and Area E (Cape Ommaney Alaska) had 71 hauls made (Table 8).

The number of species found in the catches made in Alaska (Area D and E) was nine compared to those found in the southern areas, which were sixteen in Area A, eleven in B, and fourteen in C. POP, the target species, was taken in 117 of the 127 hauls made on the trip, or 92 percent of the hauls. It was the highest, followed by *S. brevispinis* at 48 percent, *S. paucispinis* at 45 percent, *S. rubrivinctus* at 37 percent, and two species at 15 percent each—*S. ciliatus* and *S. pinniger*. The rest of the species fell below 15 percent of the hauls.

In Cape Ommaney (Area E), the skipper explored the area and found good catches of POP. There were 71 hauls made along the top of the continental slope off Cape Ommaney at depths of 100 to 143 fathoms, which was 56 percent of the hauls made during the trip. The catch rates were high for POP and held for twenty-one days. Once the catch rate of POP dropped, the vessel moved back south to Washington. While en route to and from there, she fished in Queen Charlotte Sound, along Vancouver Island, and then along the coast of Washington, where the rates were lower. There were an additional 56 hauls made in Areas A, B, C, and D on the trip to and from Area E, making a grand total of 127 hauls for Trip 3.

Table 8 *Seafreeze Pacific*—Rockfish Species Taken per Haul (Trip 3).

AREA	A	B	C	D	E	TOTAL	%
No. of Hauls	20	7	24	5	71	127	
S. aleutianus	2	–	6	–	1	9	7.08
S. alutus, POP	17	7	23	4	66	117	92.12
S. brevispinis*	10	5	13	2	31	61	48.03
S. ciliatus*	–	–	–	3	16	19	14.96
S. crameri	6	1	2	–	2	11	8.66
S. diploproa	3	3	1	–	–	7	5.51
S. elongatus	3	2	–	–	–	5	3.9
S. entomelas*	1	–	1	–	–	2	1.57
S. flavidus*	4	–	4	–	–	8	6.30
S. helvomaculatus	5	4	1	1	–	11	8.66
S. melanostomus	1	–	–	–	–	1	.78
S. paucispinis	17	7	21	1	11	57	44.88
S. pinniger	5	1	11	–	2	19	14.96
S. proriger	–	–	5	1	5	11	8.66
S. reedi	6	–	–	–	–	6	4.72
S. ruberrimus	3	2	6	–	–	11	8.66
S. rubrivinctus	7	7	21	3	9	47	37.00
S. saxicola	–	–	–	1	–	1	.78
S. zacentrus	7	2	2	2	–	13	10.24
No. of Species	16	11	14	9	9		

* = Black Rockfish

It is known that as you go north, the species of rockfish start declining. Apparently, this occurred here during Trip 3, where sixteen species were found in the hauls off Washington (Area A), eleven to fourteen off British Columbia (Areas B and C), and nine off Alaska (Areas D and E).

Observers assigned to the *Seafreeze Pacific* were told to look for and record any salmon and halibut that were taken. There were 84 salmon taken on Trip 1, 18 on Trip 2, 2 on Trip 3, 20 on the first half of Trip 4, and 2 on the last half of Trip 4. There were an estimated 178 salmon taken primarily off the Washington coast during Trip 5. Halibut were also counted and recorded by the observer when they were on board. There were 233 on Trip 1, 167 on Trip 2—of which 66 were tagged and released alive—405 on Trip 3, and 993 on the first half of Trip 4. There were an additional 668 halibut recorded for the second half of Trip 4 and 30 recorded during Trip 5 when there was no observer aboard.

The *Seafreeze Pacific* started fishing on her first trip on November 14, 1969, the year the Soviet fleet POP catch rates must have fallen. I believe that they did not target POP after that date and switched to hake. The catch of POP from the five trips of the *Seafreeze Pacific* was taken from the continental slope of 100 to 180 fathoms from the coast of Washington State to Kodiak, Alaska. The total catch of a little more than 3 million pounds was landed in Bellingham, Washington. There must have been an accounting of the landings in the state of Washington. I'm not sure how many Soviet stern trawlers there were in 1969, but in 1968, there were between ten and forty off the Washington and Oregon coasts. You can see the amount of effort that can be put in one area where fish are found by a single stern trawler, such as the *Seafreeze Pacific* did in the seventy-one hauls in the Cape Ommaney, AK, area in 1970. You can imagine what ten to forty Soviet stern trawlers could do when fish were found in 1968, the last year the Soviets concentrated on POP. During the five trips of the *Seafreeze Pacific*, there were only a few Soviet stern trawlers, and they were fishing for hake. I do not think any other skipper could have done as well as Captain Erling Jacobsen did in capturing POP in a collapsed fisheries that the Soviets were abandoning.

Fig. 41 R/V *Seafreeze Pacific*—Winter Storm Off the Washington Coast. Image #5572.

Winter Storms — *Seafreeze Pacific*

I have wondered what it was like to get caught off the Washington and British Columbia coasts in a winter storm. Fortunately, I never encountered one while I was at sea on the *Cobb*. My chances were good that I would finally be at sea during one while on the *Seafreeze Pacific* trip between November 14 and December 14, 1969. There were three periods of storms during the trip that I consider winter storms.

The first one happened at night during the beginning of the trip in Queen Charlotte Sound, a large body of water. We were facing southwest toward the Pacific Ocean from the vast opening of the sound. We were facing the oncoming waves with just enough power to hold our position. I was up in the pilothouse with the skipper, looking through the windows at the breathtaking waves as if I were watching a TV screen. Reality came when the vessel hung up on one of the waves as it passed under us. The bow and stern were not supported. The whole vessel began to vibrate bow to stern, like what I had felt when I crossed the Atlantic on a troopship during a hurricane, a feeling I had never wanted to feel again.

While we were in the pilothouse, the skipper told me he had been a captain of small trawlers, such as the *Soupfin*, and was friends with the captains of the vessels he was comparing catches with in Queen Charlotte Sound. He knew what they were going through as they tried to keep on the trawl grounds during the storm. Once it passed, they could start fishing right away, whereas if they had sought shelter, it would have been hours getting back. He said the skipper would be wedged in the pilothouse, and the vessel would be shedding water off her deck as they faced each wave as we did. These were miserable circumstances, especially on a small vessel, such as the *Soupfin*. He knew that the rest of the trawl fleet would all be tuned to one radio frequency, so he made contact with the *Soupfin's* skipper. In a joking manner, he indicated that he was just lounging in the pilothouse in his pajamas and was getting ready for bed while watching the waves go by. That is why I never wanted to be on the *Cobb*, a much smaller vessel than the *Seafreeze Pacific,* during a winter storm. I was now in a warm pilothouse, listening to the skipper's sea stories while waiting for the storm to pass.

During the second storm, the *Seafreeze Pacific* was off Vancouver Island heading into the oncoming storm waves off Barkley Sound, with just enough power to hold the position for the next morning where the captain wanted to trawl if the weather moderated. The waves were large and coming from the south. I was down in the crew's mess watching a movie, *The Heroes of Telemark*—a WWII movie about the Norwegian underground trying to stop the German army from moving heavy water to Germany, which was needed to develop an atomic bomb. We were in the part where the ferry was crossing a lake carrying the cargo. The Norwegian underground blew up the ferry, and it was sinking. The *Seafreeze Pacific* suddenly made a radical turn, which put the vessel in the wave's trough, causing her to roll. Once she got back to her original course and the movie ended, I made my way up to the pilothouse and asked what caused the violent movement. The skipper said that there was a large ship headed for *Seafreeze Pacific* from the west. She just kept coming, going east toward shore. We had to get out of the way.

The next morning, I came up to the bridge. The skipper said that the freighter that passed us the night before had crashed onto the rocks off Barkley Sound. He asked me, "Did I say I hope he goes onto the rocks?" I said, "No. You said, 'If he holds that course, he will go onto the rocks.'" A year later, I ran into the skipper, Captain Jacobsen, who testified in the trial about the loss of the freighter that passed us that night. The officer on watch on the freighter was following a loran line on his loran set for the Strait of Juan de Fuca. He thought the radar image of Barkley Sound was the Strait of Juan de Fuca. Unfortunately, he was following an echo of the correct line on the receiver, which headed him into Barkley Sound. The *Seafreeze Pacific* was holding its position, and they thought we were the *Swiftsure* lightship, which marked the entrance to the Strait of Juan de Fuca.

Toward the end of the trip, we encountered the third storm. The vessel was off the central Washington coast in a violent winter storm. The waves were huge, and the winds were strong. There was lightning during the night. In the morning, the clouds cleared, but the wind kept howling. A few of us went out on the upper deck just above the winches. The house gave us a lee, so we were not facing the wind. We were looking aft and could see the waves marching away from us toward the horizon in the distance. It was spectacular. The bow would rise up as she met the oncoming wave, and the stern was pushed down into the trough. Seawater would fill in the void of the stern ramp, which was shoved below the water level, and would quickly fill with seawater roaring up toward the deck doors at the top of the ramp. The rushing water would hit the doors and be forced up. The wind would catch it and blow it back aft. A spectacular sight (Fig. 42).

As an amateur, I wanted to determine the height of the waves we were experiencing. That is the distance from the bottom of the trough to the top of the crest. From where I was standing on the upper deck behind the house, the top of the after A-frame would go just below the horizon (Fig. 42). Then it would start up as the wave passed, presumably the bottom of the wave. Standing in the same position, when the sea was calm, the horizon would appear at the top of the stern rail. The distance from the top of the A-frame down to the top of the handrail attached to the A-frame was 25 feet. I judged that the height of the wave would be placed in code seven, 20 to 30 foot high. There is a code for wave heights from code zero (glassy calm) to code nine (phenomenal over 45 feet).[7] I felt the wave height was at least 30 feet high because the vessel seemed to be standing on its end.

Fig. 42 R/V *Seafreeze Pacific*—Wave Breaking on the Stern Ramp Doors. Image #5743.

Chapter 12: Endnotes

1. Kevin M. Bailey, *Billion-Dollar Fish: The Untold Story of Alaska Pollock* (Chicago: University of Chicago Press, 2013), pp. 93-94.

2. H. Kannt, "Boat-Types—Discussion, Comments by Some of the Authors," in *Fishing Boats of the World*, Jan-Olof Truang, ed. (London: Arthur J. Heighway Publications LTD, 1955), pp. 246.

3. H. Kannt, "German Fishing Vessels," in *Fishing Boats of the World*, Jan-Olof Truang, ed., pp. 166–170.

4. Basil Parkes, "The Owner's Viewpoint," in *Fishing Boats of the World*, Jan-Olof Truang, ed., pp. 122–126.

5. Boris O. Knake, "Operation of North Atlantic Type Otter Trawl Gear," *Fishery Leaflet* 445 (May 1958).

6. The *John N. Cobb*, Historical Ship Files, ID 119654305, *Seafreeze Pacific* reports, Seattle National Archives Office.

7. Peter Kemp, ed., *The Oxford Companion to Ships and the Sea* (Oxford: Oxford University Press, 1988), pp. 927. A code of the sea state from 0 glassy calm 0 feet to 9 phenomenal over 45 feet. Wave page 927. Similar to Beaufort Scale for winds, page 7.

CHAPTER THIRTEEN

NOAA Formed 1970

Nineteen-sixty-six was a catalytic year in my opinion that started the chage to the American fisheries. The appearance of the Soviet fleet off the Washington and Oregon coasts. The massive fishing power of that fleet devastated the POP populations off those states in four years, 1966 to 1970, caused a lot of concern, and began the change in the American fisheries. The rapid decline of the POP catch off Washington and Oregon got the politicians' attention.

The extension of the three-mile limit to twelve miles in 1967 was an attempt to save the hake fishery that was in its infant stage of development. The National Oceanic and Atmospheric Administration (NOAA) was formed in 1970. Most of the Department of the Interior's Bureau of Commercial Fisheries was transferred to the Department of Commerce's new National Marine Fisheries Service (NMFS).[1] The Soviets continued fishing hake after 1968, and they worked up an agreement with Bellingham Cold Storage, which developed into a joint venture. The Soviet Union would buy US-caught hake.

The joint venture started with a letter written to the Soviet Union's Ministry of Fisheries by Jim Talbot, the owner of Bellingham Cold Storage. He proposed that a business agreement could be made between them and the Soviet fleet fishing off the Washington coast. The letter was written after the Soviet fleet arrived off the Washington coast in 1966. They finally answered after a wait of about a year. Jim Talbot started to develop a company that would work with the Soviets. He offered a job with the new company to Dr. Wally Pereyra, who was working for NOAA in Seattle at the time. Kevin Bailey interviewed Dr. Pereyra, who explained what they had to go through to get the joint venture to work.[2] By 1977, the two-hundred-mile exclusive economic zone (EEZ) law was active, which protected our fish resources. The Soviet and US joint-venture fishery could continue working because the US vessels would catch the hake and sell them to the Soviets. They would detach the cod end of the freshly caught hake and float it over to the Soviet stern trawler standing by. The Soviets would attach a line to the cod end, pull it up their stern ramp to the deck above the factory, open the hatch, and dump the catch to factories below deck and process the hake.

The Observer Program was formed after the EEZ became law. The NMFS had the job of putting biologists aboard Russian and Japanese fishing vessels that were operating in the EEZ to record their catches. Years later, when I was wandering through a bookstore on Whidbey Island, WA, I ran across a book written by an observer, which I purchased.[3] She described her experiences on both a Japanese and a Soviet stern trawler, which was fascinating reading. I could not envision myself in the conditions she endured. I looked for more publications written by observers, which is a large program now, and learned that when they signed up, they could not write about their experiences gained there.

The *Seafreeze Atlantic* and her sister ship were laid up in 1971. The fishing industry got interested in developing its own stern trawler, so the *Seafreeze Atlantic* was purchased and modified to bottomfish in Alaskan waters. Her name was changed to the *Arctic Trawler,* and she left Seattle in May 1980. They found fish but had financial problems, so she was sold again in 1987. The factory stern trawler *Northern Glacier* was then built in Washington State in 1983. Her success encouraged the industry to turn to Norway for converting US vessels to factory stern trawlers. Between 1986 and 1990, twenty vessels were completed. The *Seafreeze Pacific* was the first US vessel to be taken to Norway for conversion in September 1985. When she was completed, her name was changed to *Royal Sea*.[4] The development of the US fishing industry having their own factory stern trawlers forced foreign fisheries vessels out of our waters.

When I returned from the *Seafreeze Pacific* trip, I became aware of rumors of coming changes in the office. I didn't pay much attention to them because I was involved in a couple of exploratory projects. The pressure to publish was off. Too much was going on. What had happened to the saying, "Go to sea and publish, and you shall be rewarded?" About this time, I remember a staff meeting where Lee Alverson was concerned about the government rules regarding raises. If you reached a certain level and didn't meet their rules, you could not advance. Higher education was one of the required rules. I'm not sure when Lee got his PhD, but sometime toward the end of the 1960s, he became Dr. Dayton L. Alverson. He certainly earned it.

In 1970, I was told to go to Lee's abandoned office down the hall and meet Ed Schaefers, who wanted to see me. I didn't know what he was doing out here in Seattle. We knew him as Uncle Ed, who was related to Uncle Sam, and he lived in Washington, DC. I came to work in 1960, just after the *Cobb* was in the shipyard getting new ribs, replaced in a dry dock because of a massive find of dry rot in the middle of the ship. She was scheduled to depart for the Chukchi Sea north of the Arctic Circle to explore the ocean floor for the AEC Project Chariot. They were planning to set off an atomic bomb to create a harbor. Lee was the new director of the exploratory base and needed emergency funds to carry out the shipyard job. As the story goes, Lee picked up the phone and called Ed, asking for additional funds to save the *Cobb*. They were granted, and the vessel was ready in time to depart as scheduled. He controlled the money, and that was why we called him Uncle Ed. I was worried. What was he doing in Seattle, and what did he want with me?

There were a lot of changes going on in our office, so perhaps he was planning to reduce the staff and was going to fire me. The last time I saw him was when he was in Seattle to review our programs. I remembered it well, and I was sure he did as well. The big man wore a serious expression as he sat in front of our group. We had been practicing our talks the night before, and I was nervous. I gave my presentation, didn't hear any comments, and answered a couple of questions. At the end of the day, Al Pruter came up to me and asked if I minded using the government car to take the big man out to the airport so he could get back to Washington, DC. I said I would and got the car, then picked Ed up. He told me we'd have to hurry since it was close to departure time. I took a shortcut I knew. He didn't say much, but after a while, he asked if I knew where I was going. I said, "Yes." After taking several turns, we arrived. I can still see him running with his bags for the doors of the airport after jumping out of the car just before the flight departed, with not a word of thanks. I was sure I was on his list of incompetents.

I went down the hall and looked into Lee's old office, seeing an empty room as I went in. In the corner, far from the sight of the office door, was a desk with Ed Schaefers sitting behind it. He motioned me to sit in the chair in front of the desk, which I did. We sat in silence.

When I was at sea during the '60s, I got to know Pete Larsen, the skipper of the *Cobb*. We had time to talk while on the bridge together as the *Cobb* ran to another location. He told me that Uncle Ed was a biologist on the *Cobb* in the early days, and he loved to argue. He would always take the opposite view, no matter what the subject. Pete told me, "Watch his nose, and if you see it start to twitch, just be aware that he is ready to say something important." So now, after this long period of silence, his nose started to twitch, and I laughed, which was followed by another period of silence. His nose started to twitch again, and I laughed. We went through the routine a couple of more times. Finally, he said, "Hitz, what on earth is wrong with you?" I said, "I'm just nervous, I guess."

Ever since that day, we hit it off. He liked people who loved boats. During WWII, he served in the Merchant Marine. He started with Exploratory Fisheries in Seattle in 1951. He sailed on the early cruise of the new vessel, *John N. Cobb,* as part of the scientific party with Lee Alverson and Don Powell. By 1960, he had moved to Washington, DC, where he became part of the mysterious top echelon of Exploratory Fishing. He told me to focus on the day-to-day activities and keep out of policy, which I did. It was years before I knew that Exploratory Fisheries was abolished when NOAA was formed.

The Department of the Interior Fish and Wildlife Service's Exploratory Fishing and Gear Research group was abolished when the Department of Commerce's NOAA was formed in 1970. The *Cobb* and all the other vessels of the Fish and Wildlife Service were transferred to the newly created NOAA fleet. It was divided into two: the Atlantic Marine Center (AMC) in Norfolk, Virginia, and the Pacific Marine Center (PMC) in Seattle, Washington. The *Cobb* was transferred to the PMC and moored with other NOAA ships at the docks on Lake Union. Sometime in early 2000, there was a change in the fleet organization. It was divided into three parts with new names: Marine Operations Center—Atlantic (MOC-A), Marine Operations Center—Pacific (MOC-P), which moved to Newport, Oregon, and the Marine Operations Center—Pacific Islands (MOC-PI) based in Hawaii.

In 1970, his job changed to helping us understand the Memorandum of Agreement between the Department of the Interior and the Department of Commerce and, specifically, how it affected the merger of the ships into the NOAA fleet. There were two positions slated at NOAA's new facility at PMC on Lake Union, which called for one civilian representative on the admiral's staff and another in the operations division, where they were planning to place me. It took four years before I was transferred officially to the PMC Operations Division in 1974. Ed was never transferred to PMC and remained on the fisheries staff. The position on the admiral's staff was never filled. It worked out since I had contact with Ed and attended the admiral's staff meetings.

My new office was in the PMC building on the pier, where the fleet tied up when in port. This was a major change for me and meant that my involvement with actual field biology at sea was at an end. Now, my shore job was involved with the management of the fisheries' vessels.

It was an interesting period of time, especially during the first few years. I remember a couple of incidents as we became used to the military environment at PMC. Ed wanted to know more about how the fishery vessels were managed after the fleet was formed. He wanted me to arrange a meeting where he could get to know my new boss, who was a captain in the NOAA Corps and head of the operations division at PMC. He had just come off one of the NOAA ships and was assigned to this job for two years, when he would be replaced by another officer. Ed also wanted to meet the head of the engineering department, who was a civilian and in the position until he retired. We met for coffee in the galley of the deactivated fisheries ship, the 214-foot *Miller Freeman*. She was deactivated just before the merger and moved to PMC. There was one individual on the vessel, an engineer. He made sure the vessel was ready to get underway. He would start the main engine and the axillary engines once in a while. The galley was a wonderful place to have the meeting because it was free of all the activities around the office.

I picked Ed up at the fisheries lab in a government vehicle, and we drove to the PMC and parked the car under the admiral's office window. I got out and was about ready to shut the car door when I looked at my free hand holding the credit card that was attached to the key, but the key was missing. I said, "What happened to the key? It's not here!" He said,

"Where is it?" We looked to see if it was in the ignition. It was not. We looked under the seat alongside it. I got on my hands and knees and looked under the car. Where was it? Were we stuck here, looking like the start of a carnival act? We finally found it in the back seat. It apparently flew off when I took it out of the ignition and somehow landed there. As we walked toward the *Miller Freeman's* gangway, Ed said, "Bob, you will never make it to be the admiral's driver! Hope he wasn't looking out the window."

We went to the galley, and the three individuals we were meeting were already there. I introduced everyone, and we sat down at a table. Ed was at the head of it, on his left was the head of the engineering department, to his right was my new boss, and I sat next to him. The ship's engineer had disappeared.

We discussed a little about the merger, and then there was silence. I got nervous. The longer the silence went on, the more nervous I was. Finally, I had to open my mouth to start them talking again, which got a little response. Then we went through the same process again, a third time, and finally, on the fourth, Ed said loudly, "Hitz, just shut up!" I finally realized that Ed wanted the other two people to get nervous and start talking, not me. Apparently, I am a slow learner.

The other incident I remember in the first few years at the PMC was about the *George B. Kelez,* a 176-foot converted navy vessel used by the fisheries for gillnetting salmon in the Gulf of Alaska to determine where the sockeye salmon go when they leave the Bering Sea—a one-purpose vessel. When she was moved to the NOAA fleet in her new berth at the PMC, she was scheduled to go into the shipyard for her annual maintenance and a minor upgrade.

I knew Tom Dunatov, the first mate on the *Kelez*, since his dad was the lead fisherman on the *Cobb* when I was aboard her. Tom had begun his career about the same time I did. He was hired as part of the crew on the *Kelez* and worked his way up to the mates' position, educating himself by getting a Coast Guard approved Merchant Marine License of a certain tonnage to cover the *Kelez*. The government didn't require the civilian officers to have a license, but the Bureau of Commercial Fisheries did, and it was carried on by NOAA when the fleet was transferred over.

The *Kelez* was acquired by the Fish and Wildlife Service when they needed a ship that could gillnet salmon in the Gulf of Alaska during the winter months. They needed to determine where the Bristol Bay sockeye salmon went after they left the freshwater of the rivers where they were born and the lakes where they remained a year before migrating to the vast Pacific Ocean. She was ideal for fishing gillnets and was an excellent seaboat. To hear stories of the blows they went through during the winter, days of one-hundred-knot winds and huge seas, the icing conditions they encountered, would raise the hair on the back of my neck.

The *Kelez* went through her shipyard period with some major items completed. Money allocated to the PMC didn't come through, so the PMC was forced to lay up ships. Since the *Cobb* and *Oregon* hadn't gone through their shipyard period and were the smallest, they were selected for deactivation. Ed Schaefers, who should have been on the admiral's staff along with Lee Alverson, said, "No!" They wanted to keep them since they were combination vessels that could fish any fishing gear, while the *Kelez*, a one-purpose vessel, should be decommissioned, even if she had already gone through the upgrade. They reversed their selection and decommissioned the *Kelez,* which was deactivated and taken out of service. Tom Dunatov was out of a job.

The reason I remember this so well was because it was near Christmas, and I was sitting at my desk at the PMC talking to Ed Schaefers when Tom and his young son Paul came in. Paul marched up to me and said loudly, "Who fired my dad?!" Ed pointed at me behind my back, and Paul, a beginning soccer player, kicked me hard on the shin. I pulled my legs up so he couldn't do it again and excitedly said, "It was Uncle Sam, not me!"

Tom said that on the way over, Paul wanted a fire truck, and Tom said there wouldn't be one that year since he was fired. Fortunately, Tom was hired back shortly afterward as the mate of the *John N. Cobb,* and finally, he became her skipper when Pete Larson retired. When Tom retired, I believe he was the last civilian skipper in the NOAA fleet.

I wish I had taken my mother's advice to keep a diary, especially during my ten years at the PMC, but I didn't. It was an interesting experience since there was a difference in manning the bridges of the two groups of vessels. Fisheries vessels were manned by civilian masters and mates, whereas other vessels were manned by corps officers from one of the seven uniformed services, such as the US Navy and Coast Guard. The difference between the groups was that the masters and mates would remain in the job until they were eligible for retirement, working up through the ranks by obtaining their mates and master licenses from the Coast Guard. Meanwhile, the NOAA Corps officer must have a college degree and complete a nineteen-week training class. Once through, they were assigned to a ship as an ensign for about two years, then two years ashore, and then back to the ship, repeating it time after time until they commanded a ship. With time, the masters and mates would retire or change jobs. When that occurred in the newly formed NOAA fleet, jobs would be filled with a corps officer. You can imagine the reactions of the civilians when they were transferred to the NOAA fleet in 1970. My job at the PMC was to make sure the fishing master and mates had a voice in this merger, but I felt that Tom and Uncle Ed had much louder voices than I did.

I remember standing on the PMC pier, watching the *Cobb* back out between the larger vessels to commence another cruise. Backing her up was not easy since she was a single screwed vessel, which did not go in a straight line when backing up but made a curve to port. Pete had to stop partway out, put the engine in forward, and swing the *Cobb's* bow to port,

Fig. 43 R/V *John N. Cobb*—Departs Pacific Marine Center (PMC). Image #5838.

which swung the stern to starboard. Then he put the vessel in reverse again to make another arc. Once out in Lake Union, he began moving forward while he swung the bow to starboard. As it swung, I saw her full profile all painted up with the large circle on the bow, the blue and white NOAA emblem. She looked elegant as she made part of a circle, showing her beautiful cruiser stern as she headed west to the ship canal, which led to the locks (Fig. 43). Once through the locks, she would enter the saltwater of the Salish Sea, where she would proceed to the Pacific Ocean. I had the feeling that I should be aboard her. How my life had changed with the assignment to the PMC and the never-ending to-do list at home.

Once the *John N. Cobb* became part of the NOAA fleet in 1970, her collected data was moved. Initially stored at the Montlake Laboratory, it was then transferred to the New Sand Point National Marine Fisheries Service (NMFS) and stored at the Resource Assessment and Conservation Engineering (RACE) Division. The *Cobb* data remained there until 2018, when it was accepted by the National Archives of the Pacific Northwest Region in Seattle, WA. It is now cataloged as Historical Ship Data Files, Records of the *John N. Cobb* (National Archives ID: 119654305) and made available to the public.

When Lee autographed my copy of his book *Race to the Sea,* he summed up this period of my career: "To Bob Hitz—one of the great exploratory scientists who served back when it was fun—Dayton 'Lee' Alverson." It was certainly an exciting time, a key time in the history of fisheries.

Chapter 13: Endnotes

1. Kevin M. Bailey, *Billion-Dollar Fish: The Untold Story of Alaska Pollock* (Chicago: University of Chicago Press, 2013), pp. 43. The Foundation of NOAA.

2. Kevin M. Bailey, *Billion-Dollar Fish*, pp. 61–66. Joint Ventures.

3. Neva Dail Bridges, *Ichthyology and Otoliths: My Life as a Foreign Fisheries Observer* (Bloomington, IN: Xlibris Publishing Company, 2004).

4. James Strong and Keith R. Criddle, *Fishing for Pollock in a Sea of Change: A Historical Analysis of the Bering Sea Pollock Fishery* (Fairbanks, AK: University of Alaska Fairbanks, 2013), pp. 24, 30–38. *Seafreeze Atlantic* on page 24, and Domestic At-Sea Production on pages 30–38.

Acknowledgments

The author would like to acknowledge the following individuals who made this book possible. Dr. Alan DeLacy, my advisor during my graduate work at UW. Dr. Dayton Lee Alverson and Al Pruter, my mentors during my employment in the exploratory group. Norm Parks, a coworker during the exploratory group when NOAA was formed who kept track of the exploratory files and kept them up to date when they were transferred to NOAA's RACE division.

After I retired from NOAA in 1988, I became interested in Carmel Finley's historical blog (https://carmelfinley.wordpress.com/about-bob/), and I wrote a posting, which she posted in 2012. She gave me confidence in my writing and encouraged me to write the book. My wife, Carolyn Maureen Hitz, edited my postings and encouraged me in my writing. Julie Pearce helped me find my way in the NOAA Sand Point facility and locate the exploratory records. She made arrangements for me to visit the facilities so I could get access through the front gate and into the building. She also helped me locate and organized the NOAA personnel to help move the records to the National Archives at Seattle on July 31, 2018.

I met with Danielle Boucher, archives specialist, to develop the Historical Ship Data Files (Records of the *John N. Cobb*), and received permission to develop and make arrangements to transfer the exploratory records to the National Archives in Seattle. Valerie Szwaya was my contact during the time I volunteered at the Archives while I worked on the *John N. Cobb* files. Another volunteer, a retired NOAA employee, Dan Twohig, became interested in the photos and negatives transferred to the Archives with the *Cobb's* information. He established a system to locate the individual photos.

My son Charles William Hitz has Sitka 2 publishing. He contracted TLC Book Design to produce my book. I am deeply indebted to him and the personnel at TLC Book Design.

APPENDIX 1

Rockfish Names

Scientific name	Common name
S. aleutianus	Rougheye Rockfish
S. alutus	Pacific Ocean Perch (POP)
S. aurora	Aurora Rockfish
S. brevispinis	Silvergray Rockfish
S. caurinus	Copper Rockfish
S. crameri	Darkblotched Rockfish
S. diploproa	Splitnose Rockfish
S. elongatus	Greenstriped Rockfish
S. entomelas	Widow Rockfish
S. flavidus	Yellowtail Rockfish
S. helvomaculatus	Rosethorn Rockfish
S. maliger	Quillback Rockfish
S. melanops	Black Rockfish
S. melanostomus	Blackgill Rockfish
S. mystinus	Blue Rockfish
S. nebulosus	China Rockfish
S. paucispinis	Bocaccio
S. pinniger	Canary Rockfish
S. proriger	Redstripe Rockfish
S. reedi	Yellowmouth Rockfish
S. rosaceus	Rosy Rockfish
S. ruberrimus	Yelloweye Rockfish (Red Snapper)
S. rubrivinctus	Flag Rockfish
S. saxicola	Stripetail Rockfish
S. wilsoni	Pygmy Rockfish

APPENDIX 2

Cruise 46, Area 2 & 3, Successful Hauls—Rockfish Species in Pounds

APPENDIX 2: Cruise 46, Area 2 & 3, Successful Hauls — Rockfish Species in Pounds

Area	2	2	2	2	2	2	2	2	2	3	3	3
Haul No.	3	4	5	6	7	8	9	10	11	12	13	15
Date	5/13/60	5/13/60	5/13/60	5/14/60	5/15/60	5/15/60	5/18/60	5/18/60	5/18/60	5/28/60	5/28/60	5/29/60
Depth	80-76	80-66	79-69	84-75	80-65	70-78	76-70	71-66	73-66	72-59	72-57	65-61
Ave. Depth	78	73	74	79.5	72.5	74	73	68.5	69.5	65.5	64.5	63
Continental	Shelf	Shelf	Shelf	Shelf	Shelf	Shelf	Shelf	Shelf	Shelf	Shelf	Shelf	Shelf
Tow Min.	60	60	60	83	110	92	30	30	46	75	80	62
Total Catch	910	1285	2000	2375	570	2900	685	400	410	5550	4540	380
BLACK R.												
S. brevispinis	250	60	-	480	200	165	485	56	300	100	25	-
S. flavidus	-	-	140	220	TR	60	22	-	TR	-	TR	TR
RED R.												
POP (S.alutus)	-	-	-	-	-	-	-	-	-	-	-	-
S. elongatus	-	-	-	TR	-	-	-	-	-	-	TR	-
S. paucispinis	TR	-	150	150	TR	45	20	-	-	100	45	TR
S. pinniger	60	-	-	60	TR	TR	TR	-	-	100	25	TR
S. rubrivinctus	-	-	-	80	-	-	-	-	-	100	185	-
S. wilsoni	-	-	-	TR	-	-	-	-	-	-	-	-

S.= *Sebastodes* or *Sebastes* R. = Rockfish TR = Less than 20 lb.

APPENDIX 3

Cruise 46, Area 4, Successful Hauls—Rockfish Species in Pounds

Area	4	4	4	4	4	4	4
Haul No.	16	18	19	20	21	22	23
Date	5/31/60	6/2/60	6/2/60	6/2/60	6/2/60	6/3/60	6/3/60
Depth	72-76	74	74-80	70-75	75-76	72-75	72-80
Ave. Depth	74	74	77	72.5	75.5	73.5	76
Continental	Shelf	Shelf	Shelf	Shelf	Shelf	Shelf	Shelf
Tow Min.	60	20	66	90	145	100	90
Total Catch	5100	475	3700	2140	1975	915	2000
BLACK R.							
S. brevispinis	1000	30	1900	1170	465	590	-
S. flavidus	-	-	-	65	-	-	-
RED R.							
POP (*S. alutus*)	250	-	-	TR	330	75	35
S. elongatus	-	-	-	-	TR	-	-
S. paucispinis	-	-	55	35	-	-	-
S. pinniger	1000	20	1335	150	200	20	20
S. proriger	2500	-	-	-	-	-	-
S. ruberrimus	-	-	-	-	TR	-	TR
S. rubrivinctus	-	-	60	-	-	-	TR

S.= *Sebastodes* or *Sebastes*   R. = Rockfish   TR = Less than 20 lb.

APPENDIX 4
Cruise 46, Area 5, Successful Hauls—Rockfish Species in Pounds

APPENDIX 4: Cruise 46, Area 5, Successful Hauls—Rockfish Species in Pounds

Area	5	5	5	5	5	5	5	5	5	5	5	5	5	5
Haul No.	24	25	26	27	28	29	30	31	33	34	35	36	37
Date	6/18/60	6/18/60	6/19/60	6/19/60	6/19/60	6/19/60	6/21/60	6/21/60	6/21/60	6/21/60	6/22/60	6/22/60	6/22/60
Depth	64-62	63-61	74-68	70-72	79-75	79-75	83-79	81-79	87-82	83-82	92-87	89-92	86-90
Ave. Depth	63	62	71	71	77	77	81	81.5	84.5	85	89.5	90.5	88
Continental	Shelf	Shelf	Shelf	Shelf	Shelf	Shelf	Shelf	Shelf	Shelf	Shelf	Shelf	Shelf	Shelf
Tow Min.	60	120	92	95	120	80	90	90	90	120	90	90	90
Total Catch	2000	3125	620	1350	5020	5010	2250	1100	5850	1540	925	2040	3700
BLACK R.													
S. brevispinis	35	175	30	135	20	TR	TR	36	200	40	30	200	50
S. flavidus	-	TR	30	56	-	-	TR	-	-	TR	TR	TR	-
S. melinops	-	-	-	-	-	-	-	TR	-	-	TR	TR	-
RED R.													
POP (S. alutus)	-	-	200	TR	5000	5000	1800	550	4000	1000	500	900	3500
S. elongates	-	-	-	TR	-	-	-	-	-	50	-	TR	TR
S. paucispinis	70	-	20	TR	-	-	TR	TR	-	-	37	225	25
S. pinniger	-	20	TR	65	TR	200	25	25	400	300	120	370	100
S. proriger	-	-	-	-	-	-	TR	-	-	-	TR	75	-
S. ruberrimus	-	TR	20	-	-	-	-	-	-	TR	-	-	-
S. rubrivinctus	TR	-	-	250	-	200	25	216	1000	100	TR	84	-

S. = Sebastodes or Sebastes R. = Rockfish TR = Less than 20 lb.

APPENDIX 5

Cruise 47, First Half, Successful Hauls — Rockfish Species in Pounds

Haul No.	39	40	41	42	45
Date	7/22/60	7/26/60	7/26/60	7/27/60	7/17/60
Depth	58	104-105	104-105	114-116	85-86
Ave. Depth	58	104.5	104.5	115	85.5
Continental	Shelf	Slope	Slope	Slope	Shelf
Tow Min.	56	75	70	75	90
Total Catch	1640	5650	1960	2120	2090
BLACK R.					
S. brevispinis	TR	1300	600	130	70
S. flavidus	-	30	TR	65	-
S. entomelas	-	TR	TR	-	-
RED R.					
POP (*S. alutus*)	-	1000	1000	1650	500
S. elongatus	-	TR	-	-	-
S. paucispinis	100	TR	-	-	TR
S. pinniger	-	-	-	-	250
S. rubrivinctus	-	TR	-	-	140
S. saxicola	-	-	TR	50	-
S. zacentrus	-	75	50	50	-

S.= Sebastodes or Sebastes   R. = Rockfish   TR = Less than 20 lb.

APPENDIX 6

Cruise 47, Ten Successful Hauls by Depth — Rockfish Species in Pounds

Haul No.	46	47	48	49	50	51	55	52	53	54
Date	8/19/60	8/19/60	8/20/60	8/22/60	8/22/60	8/23/60	8/28/60	8/23/60	8/28/60	8/28/60
Depth	50-52	52-53	71-75	77-80	82-84	92	102-90	100-108	119-106	115-111
Ave. Depth	51	52.5	73	78	83	92	96	104	112.5	113
Continental	Shelf	Shelf	Shelf	Shelf	Shelf	Shelf	Shelf	Slope	Slope	Slope
Tow Min.	65	61	91	61	67	90	60	108	62	62
Total Catch	1440	800	4320	1530	2160	5750	2565	365	1940	4090
BLACK R.										
S. brevispinis	-	-	3500	400	1500	2500	800	300	35	100
S. flavidus	-	-	50	-	30	-	50	-	-	20
RED R.										
POP (S. alutus)	-	-	-	-	20	400	1000	800	1100	300
S. elongates	-	-	TR	TR	TR	-	-	-	-	-
S. paucispinis	-	-	-	-	TR	24	TR	-	TR	-
S. pinniger	-	-	50	-	75	1500	350	1100	TR	100
S. proriger	-	-	-	-	-	TR	40	-	-	-
S. ruberrimus	-	-	100	TR	60	150	95	700	200	3000
S. saxicola	-	-	-	-	-	-	200	-	75	100

S. = Sebastodes or Sebastes   R. = Rockfish   TR = Less than 20 lb.

APPENDIX 7
Cruise 50, Area 1, Successful Hauls by Depth — Rockfish in Pounds

Area	1	1	1	1	1	1	1	1	1	1
Haul No.	19	1	18	5	12	13	17	16	15	14
Date	5/12/61	4/28/61	5/12/61	4/28/61	5/6/61	5/6/61	5/11/61	5/11/61	5/9/61	5/7/61
Depth	84-73	92-89	114-83	101-100	118-122	149-150	157-178	181-170	176-182	180-186
Ave. Depth	78.5	90.5	98.5	100.5	120	149.5	167.5	175.5	179	183
Continental	Shelf	Shelf	Shelf	Slope	Slope	Slope	Slope	Slope	Slope	Slope
Tow Min.	90	60	35	60	90	80	90	60	90	90
Total Catch	400	270	1435	500	1830	2070	440	2600	1125	350
BLACK R.										
S. brevispinis	TR	-	270	-	-	-	-	-	-	-
S. entomelas	-	-	30	-	-	-	-	-	-	-
S. flavidus	-	TR	350	TR	-	-	-	-	-	-
RED R.										
POP (S. alutus)	-	-	262	138	944	525	40	TR	TR	28
S. crameri	-	-	-	-	-	TR	TR	170	90	131
S. diploproa	-	-	-	-	-	1053	107	180	25	24
S. elongatus	TR	TR	-	-	-	-	-	-	-	-
S. paucispinis	TR	20	-	TR	49	-	-	TR	-	-
S. pinniger	35	21	35	TR	-	-	-	-	-	-
S. rubrivinctus	-	-	TR	45	285	28	TR	65	147	40
S. saxicola	-	-	-	-	TR	-	-	-	-	-
*Idiots	-	-	-	-	TR	TR	TR	70	35	TR

S. = Sebastodes or Sebastes   R. Rockfish   TR= Less than 20 lb.   *Idiots = Sebastolobus alascanus

APPENDIX 8

Cruise 50, Area 3 & 4, Successful Hauls by Depth—Rockfish in Pounds

Area	4	3	4	3	4	3	4	3	4	3
Haul No.	24	21	26	20	23	28	25	27	22	29
Date	5/14/61	5/13/61	5/15/61	5/13/61	5/14/61	5/16/61	5/15/61	5/15/61	5/14/61	5/16/61
Depth	97-86	99-101	110-117	125-140	135-130	128-140	138-139	155-172	172-180	275-300
Ave. Depth	91.5	100	113.5	132.5	132.5	134	138.5	163.5	176	287.5
Continental	Shelf	Slope	Slope	Slope	Slope	Slope	Slope	Slope	Slope	Slope
Tow Min.	57	60	60	60	60	61	60	60	60	90
Total Catch	139	4100	3050	11765	2100	3575	2520	520	470	660
BLACK R.										
S. brevispinis	TR	260	200	100	-	-	-	-	-	-
S. entomelas	-	168	-	25	-	-	-	-	-	-
RED R.										
POP (*S. alutus*)	-	292	500	11000	1500	2000	1000	50	-	-
S. crameri	-	-	600	-	400	-	500	100	-	-
S. diploproa	-	-	-	-	-	-	500	250	190	-
S. elongates	20	65	-	-	-	-	-	-	-	-
S. paucispinis	-	48	-	-	-	TR	-	-	-	-
S. pinniger	-	1055	-	-	-	-	-	-	-	-
S. rubrivinctus	-	-	200	-	-	TR	100	50	-	-
S. saxicola	-	108	1200	300	-	-	-	-	-	-
*Idiots	-	TR	-	-	23	TR	-	-	-	30

S. = *Sebastodes* or *Sebastes*   R. Rockfish   TR = Less than 20 lb.   * Idiots = *Sebastolobus alascanus*

APPENDIX 9

Cruise 50, AEC Track Line, Successful Hauls—Rockfish in Pounds

APPENDIX 9: Cruise 50, AEC Track Line, Successful Hauls — Rockfish in Pounds

Haul No.	60	61	62	63	64	65	66	68	69	70	71	72	73
Date	6/6/61	6/7/61	6/7/61	6/7/61	6/7/61	6/8/61	6/8/61	6/9/61	6/9/61	6/12/61	6/12/61	6/12/60	6/12/61
AEC Station No.	3	4	5	6	7	8	9	11	12	13	14	15	16
Station Depth F.	100	125	150	175	200	225	250	300	325	350	375	400	425
Continental	Slope	Slope	Slope	Slope	Slope	Slope	Slope	Slope	Slope	Slope	Slope	Slope	Slope
Tow Min.	60	60	60	60	60	60	60	60	60	60	60	60	60
Total Catch	700	3700	3200	2600	2400	2600	2500	600	275	230	360	200	310
BLACK R.													
S. brevispinis	TR	TR	-	-	-	-	-	-	-	-	-	-	-
S. flavidus	TR	-	-	-	-	-	-	-	-	-	-	-	-
RED R.													
S. aleutianus	-	-	-	-	-	400	150	-	-	-	-	-	-
POP (*S. alutus*)	400	3500	3000	2000	1200	300	-	-	-	-	-	-	-
S. crameri	-	50	TR	150	150	-	-	-	-	-	-	-	-
S. diplopora	-	TR	-	300	-	-	-	-	-	-	-	-	-
S. elongates	20	-	-	-	-	-	-	-	-	-	-	-	-
S. helvomaculatus	-	TR	-	-	-	-	-	-	-	-	-	-	-
S. paucispinis	-	20	TR	-	-	-	-	-	-	-	-	-	-
S. saxicola	30	-	TR	TR	-	-	-	-	-	-	-	-	-
*Idiots	-	TR	TR	102	110	70	120	40	TR	25	16	40	40

S. = *Sebastodes* or *Sebastes*   R. = Rockfish   TR = Less than 20 lb.   *Idiots = *Sebastolobus alascanus*

References

Alverson, Dayton L. "Deep-Water Trawling Survey off the Coast of Washington (August 27-October 19, 1951)." *Commercial Fisheries Review* 13, no. 11 (November 1951).

Alverson, Dayton L. "Deep-Water Trawling Survey off the Oregon and Washington Coasts (August 25-October 3, 1952)." *Commercial Fisheries Review* 15, no. 10 (October 1953).

Alverson, Dayton L., PhD. *Race to the Sea: The Autobiography of a Marine Biologist*. Bloomington, IN: iUniverse Inc., 2008.

Alverson, Dayton L. "Study of Annual and Seasonal Bathymetric Catch Patterns for Commercially Important Groundfishes of the Pacific Northwest Coast of North America." *Pacific Marine Fisheries Commission Bulletin* 4 (1960).

Alverson, Dayton L. "The Japanese and Russian Trawl Fishery in the Bering Sea." *Western Fisheries*, (April 1960).

Alverson, Dayton L. and Arthur D. Welander. "Notes on the Scorpaenid Fishes of Washington and Adjacent Areas, with a Key for Their Identification." *Copeia* 1952, no. 3 (September 1952).

Alverson, D. L., A. T. Pruter, and L. L . Ronholt. *A Study of Demersal Fishes and Fisheries of the Northeastern Pacific Ocean*. Vancouver: University of British Columbia, 1964.

Alverson, D. L., and N. J. Wilimovsky. "Fishery Investigations of the Southeastern Chukchi Sea." In *Environment of the Cape Thompson Region, Alaska*, edited by N. J. Wilimovsky and J. N. Wolfe, 843-860. Washington, DC: US Atomic Energy Commission, 1966.

Bailey, Kevin M. *Billion-Dollar Fish: The Untold Story of Alaska Pollock*. Chicago: University of Chicago Press, 2013.

Bailey, Kevin M. *The Western Flyer: Steinbeck's Boat, the Sea of Cortez, and the Saga of Pacific Fisheries*. Chicago: The University of Chicago Press, 2015.

Bridges, Neva Dail. *Ichthyology and Otoliths: My Life as a Foreign Fisheries Observer*. Bloomington, IN: Xlibris Publishing Company, 2004.

Brubaker, Bill. *Seamount: Discovery and Exploration of Cobb Seamount*. N.p.: CreateSpace Independent Publishing Platform, 2016.

Carlson, Carl B. "*S.S. Pacific Explorer*—A Preliminary Description." *Commercial Fisheries Review* 9, no. 1 (1953).

Carlson, Carl B. "Suggestions for Operators of Tuna Receiving Ships." *Fishery Leaflet* 301, April 1948.

Cleaver, Fred C. "The Washington Otter Trawl Fishery with Reference to the Petrale Sole *(Eopsetta Jordani).*" Washington State Department of Fisheries Biological Report 49a (1949).

Clemens, W. A. and G. V. Wilby. "Fishes of the Pacific Coast of Canada." *Fisheries Research Board of Canada Bulletin* 68 (1961).

Cole, James A. *Drawing on Our History: Fishing Vessels of the Pacific Northwest and Alaska.* Seattle: Documentary Media LLC, 2013.

Crowther, H.E. "Exploratory Fishing." *Fishing Gazette,* 1949 Annual Review Number 66, no. 13 (1949).

DeLacy, A. C., C. R. Hitz, and R. L. Dryfoos. "Maturation, Gestation, and Birth of Rockfish *(Sebastodes)* from Washington and Adjacent Waters." *Fisheries Research Papers* 2(3):51-67 (1964).

"Don Powell Obituary." *Seattle Times* (Archives), October 10, 1961.

Duncan, Thomas O. "The Pacific Region of the Bureau of Commercial Fisheries." *Fish and Wildlife Circular* 108, May 1961.

Ellson, J. G., Boris Knake, and John Dassow. "Report of Alaska Exploratory Fishing Expedition Fall of 1948, To Northern Bering Sea." *Fishery Leaflet* 342, June 1949.

Ellson, J. G., and Sheldon W. Johnson. "The Exploratory Fishing Vessel John N. Cobb." *Fishery Leaflet* 385, October 1950.

Fiedler, R. H., ed. "The Alaskan King Crab." Special Number. *Fishery Market News* 4, no. 5a (May 1942).

Finley, Carmel. *All the Boats on the Ocean: How Government Subsidies Led to Global Overfishing.* Chicago, IL: University of Chicago Press, 2017.

Firth, Frank E. *The Encyclopedia of Marine Resources.* New York: Van Nostrand Reinhold Company, 1969.

Fishermen's News 5, no. 9 (September 1949).

Greenwood, Melvin R. "Bottom Trawling Explorations off Southeastern Alaska, 196-1957." *Commercial Fisheries Review* 20, no. 12 (December 1958).

Greenwood, Melvin R. "Shrimp Exploration in Central Alaskan Waters by M/V *John N. Cobb,* July–August 1958." *Commercial Fisheries Review* 21, no. 7 (1959).

Gunderson, Donald. *The Rockfish's Warning.* Seattle, WA: University Book Store Press, 2011.

Hitz, Charles R. "A Trip to Nanaimo and a Last Visit with Jergen Westrheim." *Carmel Finley* (blog), December 10, 2012. https://carmelfinley.wordpress.com/2012/12/11/a-trip-to-nanaimo-and-a-last-visit-with-jergen-westrheim/.

Hitz, Charles R. "Catalogue of the Soviet Fishing Fleet." *National Fisherman Yearbook Issue* 48, no. 13 (March 31, 1968).

Hitz, Charles R. "Field Identification of the Northeastern Pacific Rockfish *(Sebastodes).*" *Fish and Wildlife Circular* 203 (1965).

Hitz, Charles R. "Observations of a Russian Trawler." *Fishermen's News,* 2nd Issue, Vol. 21, no. 11 (June 1965).

Hitz, Charles R. "R/V Commando—College of Fisheries—Off Shore." *Carmel Finley* (blog), April 4, 2015. https://carmelfinley.wordpress.com/2015/04/04/rv-commando-collage-of-fisheries-off-shore/.

Hitz, Charles R. "Soviet Hake Fleet Keeps Up Pressure." *National Fishermen* (April 1969).

Hitz, C. R. and D. L. Alverson. "Bottom Fish Survey off the Oregon Coast, April–June 1961." *Commercial Fisheries Review* 25, no. 6 (June 1963).

Hitz, C. R., H. C. Johnson, and A. T. Pruter. "Bottom Trawling Explorations off the Washington and British Columbia Coasts, May–August 1960." *Commercial Fisheries Review* 23, no. 6 (June 1961).

Hitz, Charles R., and Warren F. Rathjen. "Bottom Trawling Surveys of the Northeastern Gulf of Alaska (Summer and Fall of 1961 and Spring of 1962)." *Commercial Fisheries Review* 27, no. 9 (1965).

Ito, Daniel H., Daniel K. Kimura, and Mark E. Wilkins. "Status and Future Prospects for the Pacific Ocean Perch Resource in Waters off Washington and Oregon as Assessed in 1986." NOAA Technical Memorandum NMFS F/NWC 113 (April 1987).

Kemp, Peter, ed. *The Oxford Companion to Ships and the Sea*. Oxford: Oxford University Press, 1988.

King, Joseph E. "Experimental Fishing Trip to Bering Sea." *Fishery Leaflet* 330 (March 1949).

Knake, Boris O. "Operation of North Atlantic Type Otter Trawl Gear." *Fishery Leaflet* 445, May 1958.

Levings, Colin. "Chiefly between Kodiak Island and Cape Spencer, Alaska—a Memoir of Life on the Motor Vessel *Western Flyer* 1962–1963 and Influences on a Career in Marine Science." *Argonauta* 33, no. 3 (Summer 2016).

Liston, John, and Charles R. Hitz. *Second Survey of the Occurrence of Parasites and Blemishes in Pacific Ocean Perch, Sebastodes Alutus, May-June 1959*. Special Scientific Report—Fisheries No. 383, July 1961.

Love, Milton S., Mary Yoklavich, and Lyman Thorsteinson. *The Rockfishes of the Northeast Pacific*. Berkley, CA: University of California Press, 2002.

McNeely, Richard L. "A Practical Depth Telemeter for Midwater Trawls." *Commercial Fisheries Review* 20, no. 9 (1958).

McNeely, Richard L. "Development of the John N. Cobb Pelagic Trawl—A Progress Report." *Commercial Fisheries Review* 25, no. 7 (July 1963).

McNeely, Richard L. "Purse Seine Revolution in Tuna Fishing." *Pacific Fisherman* 59, no. 7 (June 1961).

Nelson, Martin O., and Herbert A. Larkins. "Distribution and Biology of Pacific Hake: A Synopsis." In *Pacific Hake. Fish and Wildlife Circular* 332, March 1970.

Orr, J. W., M. A. Brown, and D. C. Baker. "Guide to Rockfishes (*Scorpaenidae*) of the Genera Sebastes, *Sebastolobus*, and *Adelosebastes* of the Northeast Pacific Ocean." NOAA Technical Memorandum NMFS-AFSC-95 (October 1998).

Pereyra, W. T., H. Heyamoto, and R. R. Simpson. "Relative Catching Efficiency of a 70-Foot Semiballoon Shrimp Trawl and a 94-Foot Eastern Fish Trawl." *Fishery Industrial Research* 4, no. 1 (1967).

Phillips, Julius B. "A Review of the Rockfishes of California (Family Scorpaenidae)." *Fish Bulletin* 104 (1957).

Powell, Donald E., and Alvin E. Peterson. "Experimental Fishing to Determine Distribution of Salmon in the North Pacific Ocean, 1955." *Special Scientific Report—Fisheries* 205 (July 1957).

Pruter, A. T. "Equipment Note No. 13—Soviet Trawlers Observed in Gulf of Alaska." *Commercial Fisheries Review* 24, no. 9 (1962).

Pruter, A. T., and D. L. Alverson, eds. *The Columbia River Estuary and Adjacent Ocean Waters: Bioenvironmental Studies.* Seattle: University of Washington Press, 1972.

Records of the *John N. Cobb*. Historical Ship Files. National Archives Seattle.

"Report of the Alaska Crab Investigation." *Fisheries Market News* Vol. 4, no. 5a, May 1942—Supplement.

Schaefers, Edward A. "The John N. Cobb's Shellfish Explorations in Certain Southeastern Alaska Waters, Spring and Fall of 1950 (A preliminary report)." *Commercial Fisheries Review* 13, no. 4 (April 1951).

Schaefers, Edward A., and Frances M. Fukuhara. "Offshore Salmon Explorations Adjacent to the Aleutian Islands, June-July 1953." *Commercial Fisheries Review* 16, no. 5 (May 1954).

Smith, O. R., and M. B. Schaefer. "Fishery Exploration in the Western Pacific (January to June, 1948, by vessels of the Pacific Exploration Company)." *Commercial Fisheries Review* 11, no. 3 (1949).

Strong, James, and Keith R. Criddle. *Fishing for Pollock in a Sea of Change: A Historical Analysis of the Bering Sea Pollock Fishery.* Fairbanks, AK: University of Alaska Fairbanks, Alaska Sea Grant, 2013.

Truong, Jan Olof, Ed. *Fishing Boats of the World.* Food and Agricultural Organization of the United Nations. London, England: Arthur J. Heighway Publications LTD, 1955.

Westrheim, S. J., and H. Tsuyuki. "Sebastodes Reedi, a New Scorpaenid Fish in the Northeast Pacific Ocean." *Journal of the Fisheries Research Board of Canada* 24(9) (September 1967).

Wigutoff, Norman B. and Carl B. Carlson. "*S.S. Pacific Explorer*: Part V—1948 Operations in the North Pacific and Bering Sea." *Fishery Leaflet* 361 (January 1950).

Image Credits & Attributions

The author and publisher would like to extend their gratitude to the following sources for their contribution of images used throughout this book.

Author's Collection

Photographs: (Book Location, Image, Image Title)

Front Cover, Color Slide, Profile of *John N. Cobb*.

Back Cover, Color Slide, View from Lighthouse.

Back Cover, Left Upper Corner, Color Slide, Bob Hitz 1960.

Back Cover, Right Upper Corner, Digitized Photo, Bob Hitz 2004.

Page 25, Ed Best Picture, Fig. 9 *Pacific Explorer* with the *Tordenskjold* and seiner alongside.

Page 156, Fig. 22, Soviet R/V *Adler*, Hitz picture.

Page 227, Digitized Photo, About the Author.

Whatcom Museum, Photo Archives—Bellingham, WA.

Photographs: (Image Num., Image Title, Fig. Num., Page Num.)

1996.10.1114, *Pacific Explorer* at Bellingham, WA. Fig. 7, Page 22.

1982.92.8640, R/V *Oregon*—Rigged as a Bait Boat. Fig. 8, Page 23.

National Archives and Records Administration—Seattle, WA.
Series: Historical Ship Data Files [Records of the John. N Cobb*], 1948-1978, HS1-431659922, National Archives Identifier 119654395.*

Photographs: (Image Num., Image Title, Fig. Num., Page Num.)

2802, Soviet Passenger Ship with Side Trawler Alongside. Fig. 26, Page 162.

3055, Soviet Pushkin Class Stern Trawler. Fig. 14, Page 52.

3302, Bathythermograph (Center of Picture). Fig. 16, Page 75.

3705, United States Exploratory Research Vessel (R/V) *John N. Cobb* (1950). Fig. 2, Page 11.

3731, Pacific Ocean Perch (POP) *Sebastes alutus*. Fig. 1, Page 10.

3763, R/V *John N. Cobb* (1960), Rigged as a Trawler. Fig. 3, Page 13.

4037, R/V *Commando*—University of Washington Research Vessel. Fig. 15, Page 53.

4461, Crossing the Columbia Bar Before a Winter Storm. Fig. 19, Page 131.

4549, Modern Soviet Mayakovski Class Stern Trawler. Fig. 28, Page 165.

4586, Washington State Piper Aztec Aircraft (left to right, Gene Dinitoa and Brad Pattie). Fig. 29, Page 168.

4590, Soviet Vessel—Albatross Gully. Fig. 18, Page 123.

4593, Soviet Tanker. Fig. 25, Page 162.

4630, Soviet Stern Trawler off the Washington Coast. Fig. 30, Page 170.

4693, Soviet Vessels Anchored off Destruction Island. Fig. 23, Page 159.

4706, Soviet Side Trawler off the Washington Coast 1966. Fig.24, Page 160.

4910, Canadian R/V *G. B. Reed*—The Outboard Cable Pulled In. Fig. 38, Page 182.

4914, Canadian R/V *G. B. Reed*—The Two Cables Placed in the Snatch Block. Fig. 39, Page 182.

4920, Canadian R/V *G. B. Reed*—Connecting the Trawl Door to the Main Trawl Cable. Fig. 37, Page 182.

4999, Canadian R/V *G. B. Reed*—The Cod End Being Lifted Aboard. Fig. 40, Page 183.

5036, Canadian R/V *G.B. Reed*. Fig. 36, Page 180.

5186, Pacific Protein Corporation with *John N. Cobb* Tied Up Alongside. Fig. 27, Page 163.

5473, R/V *Seafreeze Pacific*—The Stern View of the Ramp. Fig. 32, Page 172.

5481, R/V *Seafreeze Pacific*. Fig. 31, Page 171.

5524, R/V *Seafreeze Pacific*—Bobbins Ready to be Pulled Up Around the Container. Fig. 35, Page 179.

5555, R/V *Seafreeze Pacific*—Cod End of the Net Pulled Up the Ramp. Fig. 33, Page 175.

5559, R/V *Seafreeze Pacific*—Fish Hatch Open to Fish Bin. Fig. 34, Page 175.

5572, R/V *Seafreeze Pacific*—Winter Storm Off the Washington Coast. Fig.41, Page 189.

5743, R/V *Seafreeze Pacific*—Wave Breaking on the Stern Ramp Doors. Fig. 42, Page 192.

5838, R/V *John N. Cobb*—Departs Pacific Marine Center (PMC). Fig. 43, Page 202.

11067, Large Catch of Hake. Fig. 21, Page 154.

11943, R/V *Washington*. Fig. 10, Page 27.

11945, R/V John N Cobb—Large Catch of POP, Fig. 17, Page 97.

20002, R/V *John N. Cobb*—Keel Laid. Fig. 11, Page 33.

20014, R/V *John N. Cobb*—House and Deck Work. Fig. 12, Page 33.

20036, R/V *John N. Cobb*—Launched. Fig. 13, Page 34.

20047, R/V *John N. Cobb*—Pilot House (1950). Fig. 4, Page 17.

20054, R/V *John N. Cobb*—Chart Table (1950). Fig. 5, Page 18.

20058, R/V *John N. Cobb*—Engine Room (1950). Fig. 6, Page 19.

20688, Pacific Hake—*Merluccius productus*. Fig. 20, Page 150.

Author's Drawings

Drawings: (Draw. Num., Image Title, Page Num., Comments)

Drawing 0. *Cobb* Black Silhouette, Page 1. Drawn by C. ("Chaz") W. Hitz.

  Note: The rest of the twelve drawings were drawn by C. R. ("Bob") Hitz, all with Visual CADD and converted to a PDF file and divided into three groups: Ships Profiles, Charts, Fisheries Displays.

Ships Profiles:

Drawing 1. R/V *John N. Cobb*—Profile, Page 15. The idea of making the drawing came from finding the blueprints that are now stored under The *John N. Cobb*, Historical Ship Files, ID 119654305 at the Seattle National Archives Office.

Drawing 2. R/V *John N. Cobb*—Deck Layout, Page 16. A reduced size of the blueprint of the profile and the deck layout was found in J. G. Ellson and S. Johnson's "The Exploratory Fishing Vessel *John N. Cobb*," *Fishery Leaflet* 385 (October 1950). This was used to make Drawing 1 and 2. Both have been modified over the years.

Drawing 3. R/V *Washington's* Modifications, Page 28. The drawing was done by finding the information in the document by H. C. Hanson in "Pacific Combination Fishing Vessels," in *Fishing Boats of the World* (1955) pp 200-201. In fact, Hanson's original drawings are available at the Whatcom Museum H.C. Hanson Collections in Bellingham, Washington, including one of the modification for the *Washington*.

Drawing 7. R/V *Commando*—Configuration (1960), Page 58. The drawing was made from a picture of the profile of the vessel on the cover of *Research in Fisheries* (1959), Contribution No. 77, March 1960.

Drawing 14. R/V *John N. Cobb* Modifications (1963), Page 129. Modified drawing with new deck equipment supplied for the AEC project.

Drawing 22. Canadian R/V *G.B. Reed*, Page 181. Information taken from, *Western Fisheries,* February 1963 Vol. 65 — No. 5, page 16–20. Science goes to Sea — FRB's new *G.B. Reed* page 16–20.

Charts:

  A digitizing tablet along with the use of Visual CADD in modifying or making the drawings.
  Note: The Archives have the original working charts by Cruise # in storage.

Drawing 4. *John N. Cobb* Cruise 9 Hauls, Page 41. The original drawing on page 3 of, Alverson, Dayton L. "Deep-Water Trawling Survey off the Coast of Washington (Aug. 27-Oct.1951)." *Commercial Fisheries Review* Vol. 13, No. 11 (November 1951). MS #11.

Drawing 5. *John N. Cobb* Cruise 13 Hauls, Page 46. The original drawing on page 2, Alverson, Dayton L. "Deep-Water Trawling Survey off the Oregon and Washington Coast (August 25–October 3,1952)." *Commercial Fisheries Review* Vol. 15, No 10 (October 1953). MS #18.

Drawing 6. Map of Seattle Ship Canal and Port Orchard, Page 55. The original drawing Street Road Map of Seattle.

Drawing 8. R/V *John N. Cobb* Cruise 46 Hauls, Page 65. The original drawing on page 6, Hitz, C. R., H. C. Johnson, and A. T. Pruter. "Bottom Trawling Explorations off the Washington and British Columbia Coasts, May–August 1960." *Commercial Fisheries Review* Vol. 23, No 6. (June 1961). MS #60.

Drawing 10. R/V *John N. Cobb* Cruise 47 Hauls, Page 84. The original drawing is on page 9. Same reference as Drawing 8. MS #60.

Drawing 11. R/V *John N. Cobb* Cruise 50 (First Half) Successful Hauls, Page 95. The original drawing page 3, "Bottom Fish Survey off the Oregon Coast April-June 1961". *Commercial Fisheries Review* Vol. 25, No. 6 (June 1963). MS #89.

Drawing 12. Cruise 50—Start of AEC Stations Track Line, Page 101. The original drawing is on page 6 of the same reference as Drawing 11. MS #89.

Drawing 13. R/V *John N. Cobb*—Cruise 52 and 54: Alaska Survey, Page 117. The original drawing page 4, Hitz, C. R. Hitz and W. Rathjen, "Bottom Trawling Surveys of the Northeastern Gulf of Alaska (Summer and Fall of 1961 and Spring of 1962)." *Commercial Fisheries Review* Vol. 27, No 9 (September 1965). MS #92.

Drawing 15 Cruise AEC Stations—Cruise 53 (AEC 5) and 57 (AEC 9), Page 132. Modification of Drawing 12 Cruise 50 above. MS #89.

Drawing 17. Deep Water AEC Stations—Cruise 57 (AEC 9), Page 142. Modification of Drawing 12 Cruise 50 above. MS #89.

Fisheries Displays:

Drawing 9. *R/V John N. Cobb*—Setting the Trawl, Page 78. I would have liked to draw it in 3-D as B. O. Knake did when he described the side trawler in "Operation of North Atlantic Type Otter Trawl Gear" in *Fishery Leaflet* 445 (May 1958). He was a biologist with an extraordinary artistic technique. My 2-D version is similar with the only difference being that the *Cobb* towes form the stern not by the side.

Drawing 16. 400-Mesh Otter Trawl, Gulf Shrimp Trawl, Page 140. A description of the scope ratio as well as the two different trawls.

Drawing 18. Soviet Fleet Weekly Movements off the Washington and Oregon Coasts, Page 161. Modified the drawing by adding the year 1968 to the drawing.

Drawing 19. POP Landings from Waters off Oregon, Washington, and Vancouver Island, Page 166. Converted information from NOAA Technical Memorandum NMFS F/NWC, 113 (April 1987).

Drawing 20. R/V *Seafreeze Pacific*—Aft End of the Factory Deck, Page 174. Modifications of original drawing page 21-A, in *National Fisherman* Nov. 1968.

Drawing 21. R/V *Seafreeze Pacific*—Fish Deck, Page 177. Same as Drawing 20 above.

Afterword

I dedicate this book to Scott Collins and Savonnah Mitchell, and I would like to describe what they are doing with their careers.

Scott Collins, the son of family friends, was in middle school when his mother asked me to help him on his Bridge Building Team. He was part of the Science Team at Stanwood Middle School, which competed with other Science Olympiad Schools, and I mentored him throughout the rest of his high school. Scott decided to follow science in college, where he graduated from the University of Washington as a chemical engineer and then received his PhD from North Carolina State. He and another PhD formed the company Hoofprint Biome that will hopefully reduce the methane gas emissions given off by herds of cattle. They are changing the world!

Savonnah Mitchell, my granddaughter, is also studying the sciences. She became interested in forest ecology after reading the book *Finding the Mother Tree* by Dr. Suzanne Simard and is now pursuing her BS in environmental science at Berry College (Rome, Georgia). The college is unique as it possesses twenty-seven thousand acres of land, some of which is used to conduct ecology projects. During Savonnah's freshman year, she sought out Dr. Adrienne Ernst, a professor of ecological restoration at Berry. Dr. Ernst gave Savonnah the opportunity to work with her on writing a grant proposal to do a study conducted on Berry's campus regarding restoration of the longleaf pine *(Pinus palustris)*.

The longleaf pine requires fire to reproduce, and various sites in the large campus acreage have been planted and subject to controlled burns in order to manage and increase the tree's survival. The study proposal was to identify the flora in three different sites in which controlled burns had been conducted at different times. There is little known about the different flora following these burns, and the project would involve a semi-random method of sampling the sites. The grant proposal was accepted, and Savonnah spent five weeks in the summer between her freshman and sophomore year assessing the flora, finding over 150 species. This was an amazing opportunity to work with a knowledgeable professor and learn how to put together a study from proposal to final report. Savonnah is excited to continue her work with Dr. Ernst.

I am impressed with Berry College and Savonnah's work and am anxious to see how they analyze the data they collect. Forestry and fisheries are different, as with forestry you can stand on the ground and see the surroundings in which your subjects reside. But with fisheries, you are on top of the ocean and wonder just what is happening below the surface. It will be fascinating to follow Savonnah's progress and what methods she uses.

About the Author

Mr. Charles Robert "Bob" Hitz was born during the 1930s Great Depression and was raised in Bellingham, Washington. He and his brother were expected to follow their father into dentistry and family practice.

During Bob's formative years in Bellingham, he spent significant time exploring the Salish Sea. Fascinated by the variety of vessels he encountered—such as tugs, fishing boats, and pleasure boats—he developed a keen interest in understanding their purposes. He was especially intrigued by the fishing boats, including gill netters, trollers, seiners, and trawlers. Summers spent in the San Juan Islands further fueled his curiosity, as he became captivated by the diverse fish populations he observed during these vacations.

After completing high school, Bob enrolled at Washington State University in Pullman, Washington, to pursue a profession in the dental field as per his father's wishes. However, after three years at the university, he was drafted into the US Army and stationed in France from 1954 to

1956, at the end of the Korean conflict. Upon returning home, he completed his fourth year at Washington State University and graduated with a bachelors of science degree in zoology. Seizing the opportunity provided by the GI Bill, he pursued further education in 1958 by enrolling in the graduate program at the University of Washington's College of Fisheries in Seattle, Washington. This decision ended his family's aspirations for him to become a dentist and allowed him to follow his passion for the ocean.

In 1960, Mr. Hitz decided to leave his academic pursuits for a unique job offer from the federal government. He joined the Department of the Interior's US Fish and Wildlife Service as a marine biologist at their Exploratory Fishing and Gear Research Base, located at the Montlake Laboratory in Seattle, WA. During his tenure, he participated in numerous cruises aboard the research vessel *John N. Cobb*, conducting exploratory fishing research. Ten years later, the National Oceanic and Atmospheric Administration (NOAA) was formed in 1970. The *John N. Cobb* was transferred to NOAA's Pacific Marine Center (PMC), as part of NOAA's Pacific fleet on the eastern shore of Lake Union in Seattle. Mr. Hitz continued his dedicated service after his transfer to PMC as a specialist in fisheries research vessels.

After an illustrious thirty-year career in government service, Mr. Hitz retired in 1988. Following his retirement, he expanded his cottage business, H and H Studios (https://handhstudios.com/), which specialized in crafting ship and station plaques from his pen and ink drawings. These plaques were primarily favored by the US Coast Guard. The business thrived and was sold in 2003, continuing to operate under the same name.

Currently, Bob Hitz and his wife enjoy a comfortable retirement in Stanwood, Washington, overlooking Puget Sound, reflecting on the rewarding journey of his extensive career and the successful business venture they once operated together.